OXFORD STUDIES IN NUCLEAR PHYSICS

GENERAL EDITOR
P. E. HODGSON

OXFORD STUDIES IN NUCLEAR PHYSICS

General Editor: P. E. Hodgson

1. J. McL. Emmerson: *Symmetry principles in particle physics* (1972)
2. J. M. Irvine: *Heavy nuclei, superheavy nuclei, and neutron stars* (1975)
3. I. S. Towner: *A shell-model description of light nuclei* (1977)
4. P. E. Hodgson: *Nuclear heavy-ion reactions* (1978)
5. R. D. Lawson: *Theory of the nuclear shell model* (1980)
6. W. E. Frahn: *Diffractive processes in nuclear physics* (1985)
7. S. S. M. Wong: *Nuclear statistical spectroscopy* (1986)
8. N. Ullah: *Matrix ensembles in the many-nucleon problem* (1987)
9. A. N. Antonov, P. E. Hodgson, and I. Zh. Petkov: *Nucleon momentum and density distributions in nuclei* (1988).

NUCLEON MOMENTUM AND DENSITY DISTRIBUTIONS IN NUCLEI

A. N. ANTONOV

Institute of Nuclear Research and Nuclear Energy, Bulgarian Academy of Sciences

P. E. HODGSON

Department of Nuclear Physics, University of Oxford

AND

I. Zh. PETKOV

Institute of Nuclear Research and Nuclear Energy, Bulgarian Academy of Sciences

CLARENDON PRESS · OXFORD
1988

Oxford University Press, Walton Street, Oxford OX2 6DP
Oxford New York Toronto
Delhi Bombay Calcutta Madras Karachi
Petaling Jaya Singapore Hong Kong Tokyo
Nairobi Dar es Salaam Cape Town
Melbourne Auckland

and associated companies in
Beirut Berlin Ibadan Nicosia

Oxford is a trade mark of Oxford University Press

Published in the United States
by Oxford University Press, New York

© A. N. Antonov, P. E. Hodgson, and I. Zh. Petkov, 1988

All rights reserved, No part of this publication may be reproduced,
stored in a retrieval system, or transmitted, in any form or by any means,
electronic, mechanical, photocopying, recording, or otherwise, without
the prior permission of Oxford University Press

British Library Cataloguing in Publication Data
Antonov, A. N.
Nucleon momentum and density distributions
in nuclei.—(Oxford studies in nuclear physics).
1. Nuclear physics
I. Title II. Hodgson, P. E. III. Petkov, I. Zh.
539.7   QC776
ISBN 0-19-851726-2

Library of Congress Cataloging in Publication Data
Antonov, A. N.
Nucleon momentum and density distributions in nucei.
(Oxford studies in nuclear physics)
Bibliography: p.
Includes indexes.
1. Angular momentum (Nuclear physics) 2. Angular
distribution (Nuclear physics) 3. Nuclear structure.
I. Hodgson, P. E. (Peter Edward) II. Petkov, I. Zh.
III. Title. IV. Series.
QC793.3.A5A58 1988   539.7'2   87-21947
ISBN 0-19-851726-2

Filmset and Printed in Northern Ireland by The Universities Press (Belfast) Ltd.

PREFACE

The nucleons inside the nucleus are in ceaseless motion, and these motions have important observable consequences. For example, the threshold energy for pion production is much higher for a stationary nucleon than it is for an ensemble of nucleons with a range of momenta; some of the nucleons in the target nucleus are moving towards the incident nucleon, and this greatly facilitated the first observation of pion production by Lattes and Gardner (1948). For the same reason the energies of the outgoing protons in a (p, 2p) or (e, e′p) reaction at high energies, when the incident particle interacts principally with a single nucleon, are spread around their kinematic values for a stationary nucleon by the motion of the nucleons.

It is thus important to determine the momentum distribution of the nucleons in the nucleus, and to integrate that knowledge with what we already know of nuclear structure. Remembering that the momentum and spatial wavefunctions are Fourier transforms of each other, it might be thought that there is a simple and direct connection between the nucleon momentum and density distributions. Such a relation does exist on the basis of the exact total wavefunction. The solving of the many-body problem is, however, complicated by the presence of short-range correlations due to the characteristics of the nucleon–nucleon interaction at very short distances, which produce high-momentum components in the momentum distribution. Different approximations are made in various nuclear models used to solve the many-body problem, and this introduces further differences in the relation between the momentum and density distributions. The density distributions calculated from the Hartree–Fock or single-particle potential models do not include all the important effects of these short-range correlations, and so the corresponding high-momentum components are absent from the momentum distribution. The effects of the short-range correlations may be included for few-nucleon systems; but the calculations become prohibitively complicated for all but the smallest nuclei. It is thus not easy to calculate the momentum distribution of the nucleons in nuclei.

The nucleon momentum and spatial distributions of nuclei are thus less readily connected than might at first be thought. It is however possible to establish relations between them in ways that are discussed in this book.

The principal aim of this book is to summarize our present knowledge of the momentum distribution and its connection with the density distribution of nucleons in nuclei. To do this we first review the essential quantum-mechanical formalism that enables us to describe the distributions precisely and to calculate their observable consequences (Chapter 1). In this connection special attention is paid to the density-matrix representation as well as to the Green-function method. The natural-orbital formalism is thoroughly discussed as the necessary basis of the subsequent theoretical considerations.

In this book we are primarily concerned with the development of the main theoretical methods for the description of the nucleon momentum and density distribution in nuclear matter and of finite nuclei. The few-body systems ($A = 2, 3$) which allow a more microscopic approach are not considered.

A theoretical scheme is suggested on the basis of the Hohenberg–Kohn theorem in which density and momentum distributions enter equivalently as fundamental variables of the theory (Chapter 2). A particular example of such a theoretical model is proposed in the book.

Some qualitative considerations of the behaviour of the nucleon momentum distribution in the system with different types of interactions are given in Chapter 3, together with the results of the schematic solvable model for an N-particle system with one-dimensional Hamiltonian and delta-forces. Special attention is paid to the asymptotic behaviour of the momentum distribution and of the form factor of such a system.

The consideration of the momentum and density distributions in the independent-particle models (non-interacting Fermi-gas, nuclear shell-model, Hartree–Fock) is the main subject of Chapter 4. The inability of Hartree–Fock type approximations to reproduce simultaneously the density and momentum distributions of nuclei is discussed in the same chapter.

The subsequent four chapters are devoted to some descriptions of the nucleon momentum and density distributions using methods going beyond the Hartree–Fock theory.

Chapter 5 is devoted to the consideration of the perturbation expansion of the mass operator as well as to its hole-line expansion. In the former expansion the terms following the Hartree–Fock term are considered. In the latter case attention is paid to the Brueckner–Hartree–Fock term of the hole-line expansion of the mass operator. The considerations are related closely to the problem of the nucleon momentum distribution.

The coupled-cluster form of the many-body theory (the so-called $\exp(S)$ method) is also discussed in Chapter 5 in connection with this problem.

The Jastrow-type correlation methods together with some variational

Jastrow-type calculations of the nucleon momentum and density distributions in nuclei are considered in Chapter 6. An explicit inclusion of short-range and tensor correlations to the calculations of nucleon momentum distributions and form factors of finite nuclei is discussed as well.

In Chapter 7 some natural-orbital approaches to nuclear density and nucleon momentum distributions, such as the single-particle potential method and the model of Jaminon *et al.* are considered. In these approaches many plausible features of the single-particle description can be preserved including at the same time some important short-range and tensor correlation effects.

An approach to nucleon momentum and density distributions in the generator-coordinate method using different construction potentials and Skyrme-type forces is discussed in Chapter 8. Several phenomenological models accounting for the finite-size effects in the nucleon momentum distribution are considered also in this chapter.

The problems of the determination of the nucleon momentum distribution from the experimental data are discussed in Chapters 3 and 8. It is shown that at present it is not possible to extract this distribution directly from the available data concerning different types of nuclear reactions. Many effects such as the final-state interaction prevent the direct determination of the momentum distribution, so that a model-dependent treatment of the different nuclear processes involving the momentum distribution is unavoidable.

We hope that this book will be found useful by theoreticians who want to understand nuclear momentum and density distributions and by experimentalists who want to extract them from their data.

We thank Dr V. A. Nikolaev and Dr Chr. V. Christov of Sofia and Dr F. Malaguti, Dr A. Uguzzoni, and Dr E. Verondini of Bologna who have collaborated with us on studies of nuclear momentum and nuclear density distributions respectively. They have contributed in several important respects to the work described here and we thank them for allowing us to use material which we have published together.

We thank the Bulgarian Academy of Sciences and the Royal Society of London for supporting the exchange agreement that has made our co-operation possible.

We are also grateful to all colleagues, authors, and publishers who have permitted us to reproduce illustrations from their work.

Sofia and A. N. A.
Oxford P. E. H.
January 1987 I. Zh. P.

CONTENTS

1. **Theoretical bases for the treatment of the density and momentum distribution** — 1
   1.1. Introduction — 1
   1.2. Density matrices — 3
   1.3. Second-quantization representation — 7
   1.4. Green-function method — 10
   1.5. Natural-orbital representation — 15

2. **Local density formalism and general relation between nucleon momentum and density distributions** — 19
   2.1. Definition of the nucleon momentum and density distribution — 19
   2.2. Local density formalism — 22
   2.3. The Hohenberg–Kohn theorem and the general functional relation between the nucleon momentum and density distributions — 26

3. **General considerations related to nucleon momentum distributions** — 34
   3.1. Qualitative arguments concerning the nucleon momentum distribution — 34
   3.2. Determination of the nucleon momentum distribution from experimental data — 38
   3.3. A schematic solvable model — 41

4. **Independent-particle model description** — 44
   4.1. The Fermi-gas model — 44
   4.2. The shell-model — 48
   4.3. The Hartree–Fock approximation — 51
   4.4. A remark on the reliability of Hartree–Fock predictions — 59

5. **Beyond Hartree–Fock methods** — 60
   5.1. Perturbation expansion of the mass operator — 60
   5.2. A hole-line expansion: the Brueckner–Hartree–Fock approach — 65
   5.3. The $\exp(S)$ approach — 70

6.	**Phenomenological correlation methods**	75
	6.1. Jastrow-type methods	75
	6.2. Variational Jastrow-type calculations	83
	6.3. Short-range and tensor correlation effects in the two-body density matrix	88
7.	**Natural-orbital calculations of nuclear density and nucleon momentum distributions**	96
	7.1. Introduction	96
	7.2. The single-particle potential method: nuclear density distributions	97
	7.3. The single-particle potential method: nucleon momentum distributions	113
	7.4. Another phenomenological model	118
8.	**Other phenomenological models. Experimental data related to the nucleon momentum distribution**	122
	8.1. Finite-size effects in the nucleon momentum distribution	122
	8.2. An approach to nucleon momentum and density distributions in the generator-coordinate method	126
	8.3. The coherent density fluctuation model	134
	8.4. Comparison with experimental data	141

References   151

Author Index   159

Subject Index   163

1

THEORETICAL BASES FOR THE TREATMENT OF THE DENSITY AND MOMENTUM DISTRIBUTION

In this chapter we summarize the quantum-mechanical formalism required to describe the nuclear density and nucleon momentum distribution.

1.1 Introduction

In non-relativistic quantum mechanics many-particle systems such as molecules, atoms, and nuclei are described by the wavefunction

$$\psi(r_1, r_2, \ldots r_A), \tag{1.1}$$

which depends on $3A$ coordinates and, in principle, on additional variables such as spins and isospins. This wavefunction is a solution of the stationary Schrödinger equation in the case of bound systems:

$$\hat{H}\psi = E\psi. \tag{1.2}$$

The Hamiltonian operator $\hat{H}$ may to a good approximation be expressed in terms of two-particle interactions in the form:

$$\hat{H} = \hat{T} + \hat{V} = -\sum_{i=1}^{A} \frac{\hbar^2}{2m_i} \nabla_i^2 + \sum_{i<j} v_{ij}(r_i, r_j). \tag{1.3}$$

The eigenfunctions and eigenvalues of eqn (1.2) $\psi_0, \psi_1, \ldots$ and $E_0, E_1, \ldots$ are the wavefunctions and energies of the system's states. The lowest eigenvalue E_0 and the corresponding eigenfunction ψ_0 are the binding energy and ground-state wavefunction respectively.

The complete wavefunction has symmetry properties related to the statistics of the system. Nuclei are fermion systems so the wavefunction must be antisymmetric with respect to the permutation of any two particles:

$$\hat{P}\psi(r_1, \ldots r_A) = (-1)^p \psi(r_1, \ldots r_A), \tag{1.4}$$

where $\hat{P}$ is a permutation operator acting on the indices of the A coordinates (including spin and isospin variables) and p is its parity.

In this book we are concerned with bound systems, so that the wavefunctions satisfy the condition

$$\int |\psi_n|^2 \, d\tau < \infty \qquad (1.5)$$

and since $\hat{H}$ is a Hermitian operator the $\{\psi_n\}$ form a complete orthonormal set:

$$\int \psi_n^* \psi_m \, d\tau = \delta_{mn}, \qquad (1.6)$$

$$\sum_n \psi_n^+(r_1, \ldots r_A)\psi_n(r_1', \ldots r_A') = \delta(r_1 - r_1')\delta(r_2 - r_2')\ldots \qquad (1.7)$$

It is well known that eqn (1.2) cannot be solved for any but the smallest systems because of the technical difficulties resulting from the size of the configuration space. An additional particular difficulty arises in the case of nuclear systems due to our lack of knowledge of the nucleon–nucleon interaction.

For practical purposes, however, we do not need all the information which is contained in the complete wavefunction ψ. Most of the observables important in practice such as energy, density and all the various moments, are described by one- or two-particle operators and their averages are determined by means of mathematical expressions which are defined in a limited configuration space. To illustrate this we consider the average values of the two-body operator $\hat{O}(r_1, r_2)$:

$$\langle \psi(r_1, \ldots r_A) | \hat{O}(r_1, r_2) | \psi(r_1, \ldots r_A) \rangle. \qquad (1.8)$$

In this matrix element a direct integration of $\psi^+ \psi$ is performed over all but the r_1 and r_2 radius vectors. The average value (1.8) is thus determined by the subsequent integrations over r_1 and r_2 of the integrand which is thus a much simpler object defined in a 6-dimensional space.

In the following section we define the one-body and two-body density matrices and summarize their more important properties. In Section 3 we introduce the second-quantization representation using the quantum field-theory operators, and use them to express the one- and two-body density matrices. We then use the operators representing the creation and annihilation of particles in particular quantum states to define the Hamiltonian, and this provides the relation between the state vectors in different representations and between the field operators. We next define in Section 1.4 the one-particle Green function of a system of A interacting particles and give its relation to the density matrices and to the momentum and density distributions. This enables us to introduce the mass operator and hence the basic Dyson equation which may be solved to give the Green function. The properties of the Green function and the

THEORETICAL BASES

mass operator are summarized, together with their relation to the spectral function and momentum distribution. Finally, in Section 1.5 the diagonal form of the Green function is used to obtain the one-body density matrix in the natural-orbital representation, and we summarize the properties that make it particularly appropriate for describing the nucleon momentum and density distributions.

1.2. Density matrices

The one-body density matrix

For a system of A identical particles the one-body density matrix (Dirac 1930; Löwdin 1955) is defined in terms of the complete wavefunction $\psi(r_1, \ldots r_A)$ by:

$$\rho(r, r') = A \int dr_2\, dr_3 \ldots dr_A\, \psi^+(r', r_2, \ldots r_A)\psi(r, r_2, \ldots r_A), \quad (1.9)$$

where the integration is carried over the radius vectors and summation over spin and isospin variables is implied.

The importance of the one-body density matrix can be seen by recalling its main properties.

The expectation value of an arbitrary one-particle operator

$$\hat{O} = \sum_i \hat{O}(r_i) \quad (1.10)$$

may be expressed in terms of $\rho(r, r')$ by:

$$\langle \hat{O} \rangle = \int O(r, r')\rho(r', r)\, dr\, dr' \equiv \text{Tr}[\hat{O}\hat{\rho}], \quad (1.11)$$

where $O(r, r')$ is the matrix representation of the operator $\hat{O}$ with the particular choice of dynamical variables. In the coordinate representation we have:

$$O(r, r') = \hat{O}(r)\delta(r - r'). \quad (1.12)$$

Eqn (1.11) can be obtained as follows:

$$\langle \hat{O} \rangle \equiv \int \psi^+(r_1, \ldots r_A) \sum_i \hat{O}(r_i)$$
$$\times \psi(r_1, \ldots r_A)\, dr_1 \ldots dr_A$$
$$= A \int \psi^+(r_1, \ldots r'_j, \ldots r_A)\delta(r_j - r'_j)\hat{O}(r_j)$$
$$\times \psi(r_1, \ldots r_j, \ldots r_A)\, dr_1 \ldots dr_A\, dr'_j$$
$$= \int \hat{O}(r', r)\rho(r, r')\, dr\, dr'. \quad (1.13)$$

In this derivation eqns (1.9) and (1.12) have been used.
As an illustration, if

i) $$\hat{O}(r) = r^2, \qquad (1.14)$$

then

$$\langle \hat{O} \rangle = \int \delta(r - r') r^2 \rho(r, r') \, dr \, dr' = \int r^2 \rho(r, r) \, dr. \qquad (1.15)$$

ii) $$\hat{O}(r) = -\frac{\hbar^2}{2m} \nabla_r^2, \qquad (1.16)$$

then

$$\langle \hat{O} \rangle = -\frac{\hbar^2}{2m} \int \delta(r - r') \nabla_r^2 \rho(r, r') \, dr \, dr'$$
$$= -\frac{\hbar^2}{2m} \int [\nabla_r^2 \rho(r, r')]_{r'=r} \, dr. \qquad (1.17)$$

Of special importance are the diagonal elements of the one-body density matrix:

$$\rho(r, r) = \rho(r), \qquad (1.18)$$

$$\int dr \rho(r) = A. \qquad (1.19)$$

Note that eqn (1.19) follows from (1.11) when $\hat{O} = \hat{1}$, where $\hat{1}$ is the unit operator.

Bearing in mind the definition (1.9) the diagonal elements of $\rho(r, r')/A$ have a simple physical meaning as the probability of finding a particle at point r (with spin s and isospin τ) when all other particles have arbitrary positions and spins. $\rho(r, r) \equiv \rho(r)$ is widely known as the density distribution, or simply as the local density.

Now we give the one-body density matrix in the special and practically important case when the wavefunction $\psi(r_1, \ldots r_A)$ in (1.9) can be expressed as a single determinant depending on the single-particle wavefunctions $\varphi_i(r)$. Then

$$\rho(r, r') = \sum_{i=1}^{A} \varphi_i^*(r') \varphi_i(r), \qquad (1.20)$$

and

$$\rho(r) = \sum_{i=1}^{A} |\varphi_i(r)|^2, \qquad (1.21)$$

where $\varphi_i(r)$ are the single-particle functions entering in the determinant

$\psi(r_1, \ldots r_A)$. In this case the density matrix is idempotent:

$$\rho^2 = \rho, \tag{1.22}$$

i.e.

$$\int \rho(r, r'') \rho(r'', r') \, dr'' = \sum_{i=1}^{A} \sum_{j=1}^{A} \varphi_j^*(r') \varphi_i(r) \int dr'' \varphi_i^*(r'') \varphi_j(r'')$$

$$= \sum_{i=1}^{A} \varphi_i^*(r') \varphi_i(r) = \rho(r, r'). \tag{1.23}$$

We mention that the property (1.22) is the necessary and sufficient condition for writing the function ψ as a single determinant.

The one-body density matrix (1.9) is Hermitian so that its eigenvalues are real numbers. These properties of $\rho(r, r')$ will be considered and used extensively later on in this book.

As a simple but important example of calculating the one-body density matrix in a determinant approximation we shall find $\rho(r, r')$ for the case of free particles. Single-particle states are described by plane waves

$$\frac{1}{\Omega^{1/2}} e^{ik \cdot r}, \tag{1.24}$$

so that

$$\rho(r, r') = \frac{4}{\Omega} \sum_{k < k_F} e^{-ik \cdot r'} e^{ik \cdot r}, \tag{1.25}$$

where k_F is the Fermi-momentum

$$k_F = \left(\frac{3\pi^2}{2} \frac{A}{\Omega}\right)^{\frac{1}{3}}. \tag{1.26}$$

Here the spin and isospin degeneracy is taken into account by the factor of 4 in eqn (1.25).

For large systems when $\Omega \to \infty$ and $A \to \infty$ (but $A/\Omega \equiv \rho_0 = $ constant) the summation in (1.25) can be replaced by integration using the rule:

$$\sum_k \to \frac{\Omega}{(2\pi)^3} \int d\mathbf{k}. \tag{1.27}$$

The result for $\rho(r, r')$ in this case is

$$\rho(r, r') = 3\rho_0 \frac{j_1(k_F |r' - r|)}{k_F |r' - r|}, \tag{1.26}$$

where j_1 is the first-order spherical Bessel function

$$j_1(x) = (\sin x - x \cos x)/x^2. \tag{1.29}$$

The two-body density matrix

The two-body density matrix is defined as follows:

$$\rho(r_1, r_2; r_1', r_2') = \tfrac{1}{2}A(A-1) \int dr_3 \ldots dr_A \psi^+(r_1', r_2', r_3, \ldots r_A)$$
$$\times \psi(r_1, r_2, r_3, \ldots r_A). \quad (1.30)$$

The physical meaning of the diagonal elements $\rho(r_1, r_2; r_1, r_2)/\tfrac{1}{2}A(A-1)$ is the probability of finding simultaneously one particle in position r_1 and another in r_2.

The relation between the two-body and one-body density matrices is given by

$$\int \rho(r_1, r_2; r_1', r_2) \, dr_2 = \tfrac{1}{2}(A-1)\rho(r_1, r_1'), \quad (1.31)$$

which immediately follows from the definitions (1.30) and (1.9). The two-body density matrix is antisymmetric in the variables r_1, r_2 as well as in r_1', r_2' and is Hermitian as can be seen from (1.30). In the important practical case when $\psi(r_1, \ldots r_A)$ is a single determinant the two-body density matrix is completely determined by the one-body density matrix:

$$\rho(r_1, r_2; r_1', r_2') = \tfrac{1}{2} \sum_{i=1}^{A} \sum_{j=1}^{A} [\varphi_i(r_1)\varphi_j(r_2)\varphi_i^*(r_1')\varphi_j^*(r_2')$$
$$- \varphi_i(r_1)\varphi_j(r_2)\varphi_j^*(r_1')\varphi_i^*(r_2')]$$
$$= \tfrac{1}{2}[\rho(r_1, r_1')\rho(r_2, r_2') - \rho(r_1, r_2')\rho(r_2, r_1')]. \quad (1.32)$$

Let us emphasise that this relation holds only in the particular case when $\psi(r_1, \ldots r_A)$ is a single determinant. The diagonal elements are simply related to the local density $\rho(r)$ and the one-body matrix $\rho(r, r')$:

$$2\rho(r_1, r_2; r_1, r_2) = \rho(r_1)\rho(r_2) - |\rho(r_1, r_2)|^2. \quad (1.33)$$

For the case of free particles ($\Omega \to \infty$, $A \to \infty$, $\rho_0 = $ constant)

$$\rho(r_1, r_2; r_1, r_2) = \frac{2}{9\pi^4}k_F^6 - \frac{2}{\pi^4}k_F^6 \left[\frac{j_1(k_F|r_2 - r_1|)}{k_F|r_2 - r_1|}\right]^2, \quad (1.34)$$

where k_F is given by eqn (1.26).

Equations of motion for the one-body density

In the case of two-particle interactions with Hamiltonian (1.3), the equation of motion for $\rho(r, r')$ is of the form (Thouless 1972):

$$i\hbar \frac{\partial \rho(r, r')}{\partial t} = -\frac{\hbar^2}{2M}(\nabla^2 - \nabla'^2)\rho(r, r')$$
$$+ 2\int dr_2 [V(r - r_2) - V(r' - r_2)]\rho(r, r_2; r', r_2). \quad (1.35)$$

THEORETICAL BASES 7

This equation can be derived using the definition for $\rho(r, r')$ and the time-dependent Schrödinger equation

$$i\hbar \frac{\partial \psi}{\partial t} = \hat{H}\psi. \tag{1.36}$$

Eqn (1.35) relates the one-body density matrix to the two-body one. Similarly the two-body density matrix satisfies an equation of motion which contains the higher-order density matrices. A closed equation for the one-body density matrix alone can be obtained only in the case when $\psi(r_1, \ldots r_A)$ is a single determinant. In this case $\rho(r_1, r_2; r'_1, r_2)$ is given by eqn (1.32) and eqn (1.35) resulting in the well-known Hartree–Fock equation.

Finally we give the expression for the total energy of the system E in terms of the one- and two-body density matrices:

$$E = -\frac{\hbar^2}{2M} \int [\nabla^2 \rho(r, r')]_{r'=r} \, dr + \int V(r_1, r_2) \rho(r_1, r_2; r_1, r_2) \, dr_1 \, dr_2. \tag{1.37}$$

1.3. Second-quantization representation

Here we summarize the basic formalism of the techniques of second quantization which will be used in the following chapters of this book. In quantum field theory the field operators $\psi^+(r)$ and $\psi(r)$ are used to describe creation and annihilation of a particle at position r. In terms of these operators the Hamiltonian of the system of identical particles is written as (Abrikosov *et al.* 1963):

$$H = -\frac{\hbar^2}{2m} \int dr\, \psi^+(r) \nabla^2 \psi(r) + \tfrac{1}{2} \int dr_1 \, dr_2 \, \psi^+(r_1) \psi^+(r_2)$$
$$\times v(r_1, r_2) \psi(r_1) \psi(r_2). \tag{1.38}$$

The field operators $\psi^+(r)$ and $\psi(r)$ satisfy the following commutation relations in the case of fermions:

$$\psi^+(r)\psi(r') + \psi(r')\psi^+(r) = \delta(r - r'),$$
$$\psi(r)\psi(r') + \psi(r')\psi(r) = 0, \tag{1.39}$$
$$\psi^+(r)\psi^+(r') + \psi^+(r')\psi^+(r) = 0.$$

The operator corresponding to the particle density and the number of particles can also be written in this representation:

$$\hat{\rho}(r) = \psi^+(r)\psi(r), \tag{1.40}$$

$$A = \int \hat{\rho}(r) \, dr = \int \psi^+(r)\psi(r) \, dr. \tag{1.41}$$

Let Ψ_0 be the exact ground-state wavefunction of an A-particle system. Then the state vector

$$\psi^+(r)|\Psi_0\rangle, \tag{1.42}$$

describes the system of $A+1$ particles obtained by adding a particle in a position r.

In this representation the quantity:

$$\rho(r_1, r_2) \equiv \langle\Psi_0|\,\psi^+(r_2)\psi(r_1)\,|\Psi_0\rangle$$
$$= \langle\Psi_0|\,\hat{\rho}(r_1, r_2)\,|\Psi_0\rangle, \tag{1.43}$$

is a one-body density matrix and

$$\hat{\rho}(r_1, r_2) = \psi^+(r_2)\psi(r_1) \tag{1.44}$$

is the density-matrix operator.

Similarly the two-body density-matrix operator is given by

$$\hat{\rho}(r_1, r_2; r_1', r_2') = \tfrac{1}{2}\psi^+(r_2')\psi^+(r_1')\psi(r_1)\psi(r_2). \tag{1.45}$$

In nuclear physics it is more usual to adopt the occupation number representation. The field operators $\psi(r)$ and $\psi^+(r)$ are represented by (Abrikosov *et al.* 1963):

$$\psi(r) = \sum_\alpha \varphi_\alpha(r) a_\alpha,$$
$$\psi^+(r) = \sum_\alpha \varphi_\alpha^+(r) a_\alpha^+. \tag{1.46}$$

$\varphi_\alpha(r)$ is the wavefunction of a particle in state α; a_α^+ and a_α are operators of creation and annihilation of a particle in state α. They satisfy the following commutation relations:

$$a_\alpha^+ a_{\alpha'} + a_{\alpha'} a_\alpha^+ = \delta_{\alpha\alpha'},$$
$$a_\alpha a_{\alpha'} + a_{\alpha'} a_\alpha = a_\alpha^+ a_{\alpha'}^+ + a_{\alpha'}^+ a_\alpha^+ = 0. \tag{1.47}$$

Any one-particle or two-particle operators can be represented in terms of operators a_α^+ and a_α:

$$\hat{O}^{(1)} = \sum_{\alpha,\alpha'} \langle\alpha|\,O^{(1)}\,|\alpha'\rangle\, a_\alpha^+ a_{\alpha'}, \tag{1.48}$$

$$\hat{O}^{(2)} = \tfrac{1}{2}\sum_{\substack{\alpha_1,\alpha_2\\ \alpha_1',\alpha_2'}} \langle\alpha_1'\alpha_2'|\,O^{(2)}\,|\alpha_1\alpha_2\rangle\, a_{\alpha_1'}^+ a_{\alpha_2'}^+ a_{\alpha_2} a_{\alpha_1}, \tag{1.49}$$

where

$$\langle\alpha|\,O^{(1)}\,|\alpha'\rangle \equiv \int \varphi_\alpha^+(r)O^{(1)}\varphi_{\alpha'}(r)\,dr, \tag{1.50}$$

$$\langle\alpha_1'\alpha_2'|\,O^{(2)}\,|\alpha_1\alpha_2\rangle \equiv \int \varphi_{\alpha_1'}^+(r_1)\varphi_{\alpha_2'}^+(r_2)O^{(2)}\varphi_{\alpha_1}(r_1)\varphi_{\alpha_2}(r_2)\,dr_1\,dr_2. \tag{1.51}$$

The Hamiltonian now becomes:

$$\hat{H} = \sum_{\alpha_1 \alpha_2} \langle \alpha_2 | -\frac{\hbar^2}{2m} \nabla^2 | \alpha_1 \rangle a^+_{\alpha_2} a_{\alpha_1}$$

$$+ \tfrac{1}{2} \sum_{\substack{\alpha_1 \alpha_2 \\ \alpha'_1 \alpha'_2}} \langle \alpha'_1 \alpha'_2 | V | \alpha_1 \alpha_2 \rangle a^+_{\alpha'_1} a^+_{\alpha'_2} a_{\alpha_2} a_{\alpha_1}. \tag{1.52}$$

We now list in terms of a^+ and a operators, some quantities to be used later on. The one-body density matrix is given by

$$\rho(r_1, r_2) = \sum_{\alpha, \beta} \langle \Psi_0 | a^+_\alpha a_\beta | \Psi_0 \rangle \varphi^+_\alpha(r_1) \varphi_\beta(r_2), \tag{1.53}$$

and the particle-number operator by

$$\hat{A} = \sum_\alpha a^+_\alpha a_\alpha \equiv \sum_\alpha \hat{n}_\alpha, \tag{1.54}$$

where $\hat{n}_\alpha = a^+_\alpha a_\alpha$ is the number operator for particles in state α.

An arbitrary state vector for an A-fermion system can be written quite generally in the form

$$|\psi\rangle = \sum_{\nu_1, \ldots \nu_A} C_{\nu_1, \ldots \nu_A} a^+_{\nu_1} a^+_{\nu_2} \ldots a^+_{\nu_A} |0\rangle, \tag{1.55}$$

where the ν_i label the single-particle states and $|0\rangle$ is the vacuum, i.e. $a|0\rangle = 0$.

Up to now we have used the operators $\psi(r)$ and a which do not depend explicitly on time. The time dependence is completely contained in the wavefunction and its rate of change with time is determined by the Schrödinger equation:

$$i\hbar \frac{\partial}{\partial t} |\psi\rangle = \hat{H} |\psi\rangle. \tag{1.56}$$

This representation, in which the operators do not depend explicitly on time, is called the Schrödinger representation.

In the Heisenberg representation the wavefunctions do not depend on time and the full time dependence is contained in the operators. The time evolution of the operators is determined by the equation:

$$\frac{\partial \hat{F}_H}{\partial t} = i(\hat{H}\hat{F}_H - \hat{F}_H \hat{H}) \equiv i[\hat{H}, \hat{F}_H], \tag{1.57}$$

which immediately follows from the basic relation between the Schrödinger ($\hat{F}_S$) and the Heisenberg ($\hat{F}_H(t)$) operators

$$\hat{F}_H(t) = e^{iHt} \hat{F}_S e^{-iHt}. \tag{1.58}$$

In the interaction representation the time dependence enters both in the wavefunctions and in the operators. The relation between the operators in the interaction representation and those in the Schrödinger representation is:

$$\hat{F}_{int}(t) = e^{iH_0 t}\hat{F}_S e^{-iH_0 t}, \quad (1.59)$$

and hence

$$\frac{\partial \hat{F}_{int}}{\partial t} = i[H_0, \hat{F}_{int}], \quad (1.60)$$

where $H_0 = H - V$ and V is the part of the Hamiltonian responsible for the interaction between particles.

The relations between the state vectors in the different representations are:

$$|\psi_S(t)\rangle = e^{-iHt}|\psi_H\rangle, \quad (1.61)$$

$$|\psi_{int}(t)\rangle = e^{iH_0 t}|\psi_S(t)\rangle, \quad (1.62)$$

$$|\psi_{int}(t)\rangle = e^{-iVt}|\psi_H\rangle. \quad (1.63)$$

In all these expressions the indices S, H, int stand for the Schrödinger, Heisenberg, and interaction representations.

In the Heisenberg representation the field operator $\psi^+(\psi)$ and creation (annihilation) operators $a^+(a)$ are given by

$$\psi_H(r, t) = e^{iHt}\psi_S(r)e^{-iHt}, \quad (1.64)$$

$$a_H(t) = e^{iHt}a_S e^{-iHt}. \quad (1.65)$$

These operators will be necessary for the introduction of Green functions in the next section.

1.4. Green-function method

The one-particle Green function of A interacting particles is defined by the expression:

$$G(r, t; r', t') = -i\langle \Psi_0| T\{\psi(r, t)\psi^+(r', t')\} |\Psi_0\rangle, \quad (1.66)$$

where Ψ_0 is the exact ground-state wavefunction; $\psi(r, t)$ are field operators, both in the Heisenberg representation.

The time-ordering operator T in the case of fermions is defined by (Abrikosov *et al.* 1963):

$$T\{\psi(r, t)\psi^+(r', t')\} = \begin{cases} \psi(r, t)\psi^+(r', t'), & \text{for } t > t' \\ -\psi^+(r', t')\psi(r, t), & \text{for } t < t'. \end{cases} \quad (1.67)$$

The one-particle Green function can be used for the determination of the average value of an arbitrary one-particle operator of the type (1.10):

$$\langle \hat{O}^{(1)} \rangle = -i \int \lim_{\substack{t' \to t+0 \\ r' \to r}} O^{(1)}_{\alpha\beta}(r, t) G_{\alpha\beta}(r, t; r', t')] \, dr. \quad (1.68)$$

Here we introduce explicitly the spin and isospin indices α, β and a summation over α and β is implied. The Green function $G_{\alpha\beta}$ depends on α and β through the field operators $\psi_\alpha(r, t)$ and $\psi^+_\beta(r', t')$.

An important case is the average value of the particle-density operator $\hat{\rho}$ (1.40):

$$\rho(r) = -i \lim_{\substack{t' \to t+0 \\ r' \to r}} G_{\alpha\alpha}(r, t; r', t'). \quad (1.69)$$

The one-particle Green function is simply related with the one-body density matrix $\rho(r, r')$ (1.43):

$$\rho(r, r') = -i \lim_{t' \to t} G_{\alpha\alpha}(r, t; r', t'). \quad (1.70)$$

We shall also need the second-order (two-particle) Green function G_2:

$$G_2(r_1 t_1, r_2 t_2; r_3 t_3, r_4 t_4)$$
$$= -i \langle \Psi_0 | T\{\psi(r_1, t_1) \psi(r_2, t_2) \psi^+(r_3, t_3) \psi^+(r_4, t_4)\} | \Psi_0 \rangle. \quad (1.71)$$

The particular form of the function $G_2(r_1 t, r_2 t; r_1 t, r_2 t)$ is related to the diagonal elements of the two-body density matrix (1.30) with a proper transition to the equal times

$$G_2(r_1 t, r_2 t; r_1 t, r_2 t) \longrightarrow \rho(r_1, r_2; r_1, r_2). \quad (1.72)$$

There exists a relation between the one-particle and two-particle Green functions, which is obtained if one tries to find the time-derivative of G. The following equation can be derived:

$$\left(i \frac{\partial}{\partial t} + \frac{\nabla^2}{2m}\right) G(r, t; r', t') + i \langle \Psi_0 | T\{[\hat{V}, \psi(r, t)] \psi^+(r', t')\} | \Psi_0 \rangle$$
$$= \delta(r - r') \delta(t - t'), \quad (1.73)$$

where

$$\hat{V} = \int dr_1 \, dr_2 \psi^+(r_1, t) \psi^+(r_2, t) V(r_1, r_2) \psi(r_1, t) \psi(r_2, t), \quad (1.74)$$

and $V(r_1, r_2)$ is a two-body interaction.

The second term on the left of eqn (1.73) contains the average on the ground state Ψ_0 of four field operators. In this way a two-particle Green function G_2 is formed.

Here we are not going to discuss the relation of G_2 with higher-order Green functions, but will concentrate instead only on eqn (1.73). This equation is evidently not closed and to obtain a closed expression some additional conditions must be imposed.

We note that only in the Hartree–Fock approximation is eqn (1.73) complete, because in this case the two-particle Green function G_2 is simply related to

$$G_2(r_1t_1, r_2t_2; r_3t_3, r_4t_4) \approx G(r_1t_1; r_3t_3)G(r_2t_2; r_4t_4)$$
$$- G(r_1t_1; r_4t_4)G(r_2t_2; r_3t_3). \quad (1.75)$$

A similar relation between density matrices is given in eqn (1.32).

The exact formal treatment of eqn (1.73) is made possible by introducing the mass operator Σ by means of the equations

$$-i \langle \Psi_0| T\{[V, \psi(r, t)]\psi^+(r', t')\} |\Psi_0\rangle \equiv \Sigma G, \quad (1.76)$$

and

$$\Sigma G = \lim_{t_1 \to t+0} \int dr_1\, dr'_1\, dr'_2 \vartheta(r_1r; r'_1r'_2)G_2(r'_1t, r'_2t; r_1t_1, r'_1t'), \quad (1.77)$$

where

$$\vartheta(r_1r; r'_1r'_2) = \delta(r_1 - r'_1)\delta(r - r'_2)V(r_1, r_2). \quad (1.78)$$

Eqn (1.73) now reads:

$$\left(i\frac{\partial}{\partial t} + \frac{\nabla^2}{2m} - \Sigma\right)G = \delta(r - r')\delta(t - t'). \quad (1.79)$$

In the above-mentioned Hartree–Fock approximation Σ is equivalent to a mean self-consistent field:

$$\Sigma G = \hat{V}(r)G(r, r'). \quad (1.80)$$

Eqn (1.79) is known as the Dyson equation. It can be written in a more convenient form after the Fourier transform on the variable $\tau = t - t'$. Note that the Green functions G and G_2 depend on $t - t'$ because of time-translational invariance. Now we can write

$$\left(\omega + \frac{\nabla^2}{2m}\right)G(r, r', \omega) - \int dr_1 \Sigma(r, r_1, \omega)G(r_1, r', \omega) = \delta(r - r').$$
$$(1.81)$$

Symbolically, this equation is often expressed in operator form:

$$G = G_0 + G_0\Sigma G, \quad (1.82)$$

where G_0 is a free-particle Green function:

$$G_0 = \left(\omega + \frac{\nabla^2}{2m}\right)^{-1}. \tag{1.83}$$

Let us introduce a mixed representation of the mass operator:

$$\Sigma(\mathbf{r}, -i\nabla, \omega) = \int \Sigma(\mathbf{r}, \mathbf{r}_1, \omega)\exp[i(\mathbf{r} - \mathbf{r}_1)(-i\nabla)]\, d\mathbf{r}_1, \tag{1.84}$$

and define single-particle functions $\varphi_\alpha(\mathbf{r})$ by the equation:

$$\left[-\frac{\nabla^2}{2m} + \Sigma(\mathbf{r}, -i\nabla, \omega)\right]\varphi_\alpha(\mathbf{r}, \omega) = E_\alpha(\omega)\varphi_\alpha(\mathbf{r}, \omega). \tag{1.85}$$

The Green function $G(\mathbf{r}, \mathbf{r}, \tau = t - t')$ can be written in the α representation, which means that the field operators are expressed by means of creation and annihilation operators, see eqn (1.46)

$$G(\mathbf{r}, \mathbf{r}', \tau) = \sum_\alpha \varphi_\alpha^*(\mathbf{r})\varphi_{\alpha'}(\mathbf{r}')G_{\alpha\alpha'}(\tau = t - t'), \tag{1.86}$$

$$G_{\alpha\alpha'}(\tau) = -i\langle\Psi_0|\, T a_\alpha(t) a_{\alpha'}^+(t')\, |\Psi_0\rangle. \tag{1.87}$$

Here $\varphi_\alpha(\mathbf{r})$ are an arbitrary complete set of single-particle functions. If we use the chosen set φ_α satisfying eqn (1.85) then $G_{\alpha\alpha'}$ acquires a simple diagonal form, which in the ω representation is

$$G_{\alpha\alpha'}(\omega) = \delta_{\alpha\alpha'} G_\alpha(\omega), \tag{1.88}$$

or

$$G_\alpha(\omega) = \frac{1}{\omega - E_\alpha(\omega)}. \tag{1.89}$$

The connection between the ω and τ representations is given by

$$G_\alpha(\tau) = \int G_\alpha(\omega) e^{-i\omega\tau} \frac{d\omega}{2\pi}. \tag{1.90}$$

If ω_α is a solution of the equation

$$\omega = E_\alpha(\omega), \tag{1.91}$$

then the pole-representation of the Green function follows:

$$G_{\alpha\alpha'} = \frac{\text{Res}(\omega_\alpha)}{\omega - \omega_\alpha + i\gamma} \delta_{\alpha\alpha'}, \tag{1.92}$$

where

$$\text{Res}(\omega_\alpha) = \frac{1}{1 - \partial E_\alpha(\omega)/\partial\omega\,|_{\omega=\omega_\alpha}}. \tag{1.93}$$

The states with energies $\omega = \omega_\alpha$ are interpreted as quasiparticle excitations, the lifetime of which depends on how close this energy lies to the Fermi-energy, ω_F. The latter satisfies the equation:

$$\frac{p_F^2(r)}{2m} + \Sigma(r, p_F(r), \omega_F) = \omega_F. \tag{1.94}$$

The Fermi-momentum p_F for infinite systems does not depend on r and is related to the density of the system (1.26). Generally, the poles of the Green function are complex, i.e. the quasiparticle 'energies' have real and imaginary parts, the latter connected with the width (corresponding to the lifetime) of the quasiparticle state.

These important properties of the Green functions can be seen in a more convenient way in the case of infinite systems, where the related quantities are easier to interpret and have an unambiguous physical meaning. Space- and time-translational invariance allows the following form of the Green function in the momentum and energy representation:

$$G(k, \omega) = \frac{1}{\omega - k^2/2m - \Sigma(k, \omega)}. \tag{1.95}$$

Generally the mass operator $\Sigma(k, \omega)$ is a complex operator. It is real for $\omega \to \omega_F$ and in the vicinity of ω_F it has the following asymptotic behaviour (Hugenholtz 1957; Migdal 1967)

$$\text{Im}\,\Sigma \simeq C(\omega - \omega_F)^2. \tag{1.96}$$

The following dispersion relation exists between the real and imaginary part of the mass operator:

$$\text{Re}\,\Sigma(k, \omega) = \int_{-\infty}^{\infty} \frac{d\omega'}{2\pi} \frac{\text{Im}\,\Sigma(k, \omega')}{\omega - \omega'}. \tag{1.97}$$

Of special interest is the quantity:

$$S(k, \omega) = \frac{i}{2\pi}[G(k, \omega + i\eta) - G(k, \omega - i\eta)]$$

$$= \frac{1}{\pi} \frac{\text{Im}\,\Sigma(k, \omega)}{[\omega - k^2/2m - \text{Re}\,\Sigma(k, \omega)]^2 + [\text{Im}\,\Sigma(k, \omega)]^2}, \tag{1.98}$$

which is called the spectral function and is related to the widths of the quasiparticle states. The following relations also hold:

$$G(k, \omega) = G_h(k, \omega) + G_p(k, \omega)$$

$$= \int_{-\infty}^{\varepsilon_F} d\omega' \frac{S_h(k, \omega')}{\omega - \omega' - i\eta} + \int_{\varepsilon_F}^{\infty} d\omega' \frac{S_p(k, \omega')}{\omega - \omega' + i\eta}, \tag{1.99}$$

$$S(\mathbf{k}, \omega) = S_{\mathrm{p}}(\mathbf{k}, \omega) + S_{\mathrm{h}}(\mathbf{k}, \omega); \qquad \int_{-\infty}^{\infty} \mathrm{d}\omega S(\mathbf{k}, \omega) = 1. \qquad (1.100)$$

$S_{\mathrm{h}}(\mathbf{k}, \omega \leq \varepsilon_{\mathrm{F}})$ and $S_{\mathrm{p}}(\mathbf{k}, \omega > \varepsilon_{\mathrm{F}})$ are called hole and particle spectral functions and determine the widths of the hole and particle states respectively.

Note that the momentum distribution $n(k)$ is determined by means of the hole spectral function:

$$n(k) = \int_{-\infty}^{\varepsilon_{\mathrm{F}}} \mathrm{d}\omega' S_{\mathrm{h}}(k, \omega'). \qquad (1.101)$$

This expression will be used in Section 8.4 of this book.

1.5. Natural orbital representation

The diagonal form of the Green function in its α representation eqns (1.86), (1.88), and (1.90) gives the possibility of obtaining the one-body density matrix in the natural orbital representation (NOR). By definition

$$\rho(\mathbf{r}, \mathbf{r}') = -\mathrm{i} \lim_{t' \to t} G(\mathbf{r}, \mathbf{r}', t - t'), \qquad (1.102)$$

$$G(\mathbf{r}, \mathbf{r}'; \tau = t - t') = \sum_{\alpha \alpha'} \varphi_{\alpha}^{*}(\mathbf{r}) \varphi_{\alpha'}(\mathbf{r}') G_{\alpha \alpha'}(\tau). \qquad (1.103)$$

With the choice of the complete set of wavefunctions $\varphi_{\alpha}(\mathbf{r})$ satisfying eqn (1.85) we have

$$\rho(\mathbf{r}, \mathbf{r}') = -\mathrm{i} \lim_{\tau \to 0^{-}} \sum_{\alpha} \varphi_{\alpha}^{*}(\mathbf{r}) \varphi_{\alpha}(\mathbf{r}') G_{\alpha}(\tau), \qquad (1.104)$$

or equivalently

$$\rho(\mathbf{r}, \mathbf{r}') = \sum_{\alpha} n_{\alpha} \varphi_{\alpha}^{*}(\mathbf{r}) \varphi_{\alpha}(\mathbf{r}'). \qquad (1.105)$$

This form of the one-body density matrix was introduced by Löwdin (1955). The functions φ_{α} are the so-called natural orbitals and n_{α} is the occupation number in the state α. These numbers are related to the Green function by

$$n_{\alpha} \equiv -\mathrm{i} \lim_{\tau \to 0^{-}} G_{\alpha}(\tau), \qquad (1.106)$$

where

$$G_{\alpha}(\tau) = \int G_{\alpha}(\omega) \mathrm{e}^{-\mathrm{i}\omega\tau} \frac{\mathrm{d}\omega}{2\pi}, \qquad (1.107)$$

$$G_{\alpha}(\omega) = \frac{1}{\omega - E_{\alpha}(\omega)}, \qquad (1.108)$$

see eqn (1.89).

Using eqn (1.87) the limit in (1.106) can be obtained and written in the

form
$$n_\alpha = \langle \Psi_0| a_\alpha^+ a_\alpha |\Psi_0\rangle. \tag{1.109}$$

Now from (1.109) it follows that the values of n_α lie between 0 and 1.

In the Hartree–Fock approximation

$$n_\alpha^{HF} = -i \lim_{\tau \to 0^-} G_\alpha^{HF}(\tau) = \begin{cases} 1, & \alpha \leq \alpha_F \\ 0, & \alpha > \alpha_F. \end{cases} \tag{1.110}$$

where α_F stands for the Fermi-level.

The one-body density matrix $\rho(r_1, r_2)$ in the Hartree–Fock approximation then takes the form:

$$\rho(\boldsymbol{r}, \boldsymbol{r}') = \sum_\alpha^{\alpha_F} \varphi_\alpha^*(\boldsymbol{r}) \varphi_\alpha(\boldsymbol{r}'). \tag{1.111}$$

This form corresponds to the determinant ground-state wavefunction Ψ_0 (1.20).

In eqn (1.53) the one-body density matrix is expressed in terms of an arbitrary chosen orthonormal set of functions $\varphi_\alpha(r)$. The diagonal elements $\langle \Psi_0| a_\alpha^+ a_\alpha |\Psi_0\rangle$ (i.e. the occupation number in state α) correspond to the chosen φ representation. If one takes a sum of these numbers up to the Fermi-level (α_F), for example, the sum will in general deviate from one φ representation to another. So changing the sets of functions φ produces a rearrangement of the particles below and above the Fermi-level conserving, of course, the total number of particles. It turns out that there is a relation between the optimal arrangement of particles and the type of chosen basis. This optimal arrangement of particles is realized in the case of natural orbitals and means that the number of particles in the Fermi-sea $A_<$ is the maximal (the so-called minimal depletion of the Fermi-sea) and consequently, the number of particles above the Fermi-sea $A_>$ is minimal (Kobe 1969). This can be simply demonstrated by varying the expression

$$A_< = \sum_{\alpha,\beta}^{\alpha_F} \int d\boldsymbol{r} \varphi_\alpha^*(\boldsymbol{r}) \varphi_\beta(\boldsymbol{r}) \langle \Psi_0| a_\alpha^+ a_\beta |\Psi_0\rangle, \tag{1.112}$$

with respect to an arbitrary change of $\varphi(\boldsymbol{r})$:

$$\delta\varphi_\beta = \sum_{\gamma(\neq\beta)} \xi_{\beta\gamma} \varphi_\gamma, \tag{1.113}$$

$$\delta\varphi_\alpha^* = \sum_{\gamma(\neq\beta)} \eta_{\alpha\gamma} \varphi_\gamma^*, \tag{1.114}$$

taking into account the orthonormality condition

$$\int \varphi_\alpha^*(\boldsymbol{r}) \varphi_\beta(\boldsymbol{r}) \, d\boldsymbol{r} = \delta_{\alpha\beta}. \tag{1.115}$$

Then
$$\delta A_< = \sum_{\substack{\gamma,\beta \\ (\gamma \neq \beta)}}^{\alpha_F} \xi_{\beta\gamma} \langle \Psi_0 | a_\gamma^+ a_\beta | \Psi_0 \rangle + \text{c.c.} \tag{1.116}$$

We see that $\delta A_<$ vanishes for arbitrary variations of $\xi_{\beta\gamma}$ only if:

$$\langle \Psi_0 | a_\gamma^+ a_\beta | \Psi_0 \rangle = 0, \quad \gamma \neq \beta, \quad \gamma \leq \alpha_F, \quad \beta \leq \alpha_F. \tag{1.117}$$

Eqn (1.117) is just the condition for the natural-orbital representation of the one-body density matrix (1.105). It can be shown that the second condition $\delta^2 A_<$ is negative, so the particle number $A_<$ is maximal. The conservation of the total number of particles

$$A = A_< + A_>, \tag{1.118}$$

enables one to state that the particles above the Fermi-sea $A_>$ is minimal. Thus the natural-orbital representation corresponds to a minimal depletion of the Fermi-sea.

Since the determination of the natural orbitals and the corresponding occupation numbers is equivalent to the solution of the exact many-body problem, in practice we often deal with independent-particle model (Hartree–Fock type) approaches. It is therefore useful to have a reasonable criterion for choosing the best independent-particle model wave functions; the corresponding one-body density matrix $\rho_0(\mathbf{r}, \mathbf{r}')$ then has properties close to those of the exact one-body density matrix $\rho(\mathbf{r}, \mathbf{r}')$. A useful criterion is provided by the mean-square deviation of these two matrices (Kutzelnigg and Smith 1964)

$$\text{Tr}[(\rho - \rho_0)^2] = \text{minimum}. \tag{1.119}$$

Now we show that if ρ is the exact one-body density matrix and ρ_0 is associated with a Slater determinant built with natural orbitals then eqn (1.119) holds.

In an arbitrary φ representation of $\rho(\mathbf{r}, \mathbf{r}')$ and $\rho_0(\mathbf{r}, \mathbf{r}')$:

$$\rho(\mathbf{r}, \mathbf{r}') = \sum_{\alpha,\beta} n_{\alpha\beta} \varphi_\alpha^*(\mathbf{r}) \varphi_\beta(\mathbf{r}'), \tag{1.120}$$

$$\rho_0(\mathbf{r}, \mathbf{r}') = \sum_\alpha^{\alpha_F} \varphi_\alpha^*(\mathbf{r}) \varphi_\alpha(\mathbf{r}'), \tag{1.121}$$

with

$$n_{\alpha\beta} = \langle \Psi_0 | a_\alpha^+ a_\beta | \Psi_0 \rangle, \tag{1.122}$$

then eqn (1.119) can be written in the form (after performing the trace):

$$\sum_{\alpha,\beta} n_{\alpha\beta} n_{\beta\alpha} - 2 \sum_\alpha^{\alpha_F} n_{\alpha\alpha} + A = \text{minimum}. \tag{1.123}$$

The first and third terms of (1.123) are constants and therefore the minimum can be reached if

$$\sum_{\alpha}^{\alpha_F} n_{\alpha\alpha} \equiv A_< = \text{maximum}. \qquad (1.124)$$

As we showed above, eqn (1.124) is fulfilled if the φ's are natural orbitals.

The criterion (1.119) will be used in the analyses of nucleon density and momentum distributions in the framework of the independent-particle models.

2

LOCAL DENSITY FORMALISM AND GENERAL RELATION BETWEEN NUCLEON MOMENTUM AND DENSITY DISTRIBUTIONS

The general formalism developed in the first chapter may now be used to define the nucleon momentum and density distributions and to show how they can be related in the context of various models.

In the following section the local density and momentum distribution operators are defined and expressed in terms of the one-body density matrix, the Wigner distribution function, and the one-particle Green function. In Section 2.2 the Thomas–Fermi model is defined using the concept of quantum statistics and used to give an expression for the total energy. We also describe other phenomenological methods, such as the energy-density formalism and the extended Thomas–Fermi model, that may be used to describe the bulk properties of nuclei.

Finally we discuss the importance of the Hohenberg–Kohn theorem that shows that the many-particle wavefunction and hence all ground-state properties are unique functionals of the density. This also applies to the momentum distribution and these two distributions are treated by the theory on the same footing.

2.1. Definition of the nucleon momentum and density distribution

Here we give in a systematic way the definitions of the main quantities considered in this book, namely the nucleon density and momentum distribution and quantities related to them. The similarity of the definitions of these two quantities is of special interest bearing in mind the general intrinsic relation between $\rho(r)$ and $n(k)$ which will be illustrated later in this chapter.

The local density distribution operator $\hat{\rho}$ is written in the r and q

representations correspondingly in the forms:

$$\hat{\rho}_r = \sum_{j=1}^{A} \delta(r - r_j), \qquad (2.1)$$

$$\hat{\rho}_q = \sum_{j=1}^{A} e^{iq \cdot r_j}, \qquad (2.2)$$

where r_j are the radius-vectors of the nucleons.

Similar definitions hold for the nucleon momentum distribution operator in the q and r representations:

$$\hat{n}_q = \sum_{j=1}^{A} \delta(q - q_j), \qquad (2.3)$$

$$\hat{n}_r = \frac{1}{(2\pi)^3} \sum_{j=1}^{A} e^{ir \cdot q_j}. \qquad (2.4)$$

The observable nucleon density $\rho(r)$ and momentum $n(q)$ distributions can be defined as the expectation values of the operators (2.1) and (2.3) in the ground state represented by the many-particle wavefunction ψ in the r and q representations:

$$\rho(r) = \langle \psi_r | \hat{\rho}_r | \psi_r \rangle, \qquad (2.5)$$

$$n(q) = \langle \psi_q | \hat{n}_q | \psi_q \rangle. \qquad (2.6)$$

The expectation values of $\hat{\rho}_q$ (2.2) and $\hat{n}_r$ (2.4) calculated using the same functions ψ_r and ψ_q are the form factor $F(q)$ and the inverted form factor $\mathscr{F}(r)$:

$$F(q) = \langle \psi_r | \hat{\rho}_q | \psi_r \rangle = \int \rho(r) e^{iq \cdot r} dr, \qquad (2.7)$$

$$\mathscr{F}(r) = \langle \psi_q | \hat{n}_r | \psi_q \rangle = \int n(q) e^{iq \cdot r} \frac{dq}{(2\pi)^3}. \qquad (2.8)$$

It is useful to invert the relations (2.7) and (2.8) and express $\rho(r)$ and $n(q)$ as:

i) $\rho(r)$ by means of the form factor $F(q)$, and ii) $n(q)$ by means of the inverted form factor $\mathscr{F}(r)$:

$$\rho(r) = \int F(q) e^{-iq \cdot r} \frac{dq}{(2\pi)^3}, \qquad (2.9)$$

$$n(q) = \int \mathscr{F}(r) e^{-iq \cdot r} dr. \qquad (2.10)$$

Eqn (2.9) gives the relation between the density distribution and the experimentally measurable form factor $F(q)$.

The quantity $\mathscr{F}(r)$ in (2.10) is related to the total kinetic energy of the system T_{kin}:

$$T_{kin} = \int \frac{\hbar^2 q^2}{2m} n(q) \frac{dq}{(2\pi)^3} = -\frac{\hbar^2}{2m} \nabla^2 \mathscr{F}(r) \bigg|_{r=0}. \tag{2.11}$$

The nucleon density and momentum distributions can be expressed by the theoretical quantities, such as the one-body density matrix $\rho(r, r')$, the Wigner distribution function, and the one-particle Green function.

By definition the one-body density matrix is:

$$\rho(r, r') = A \int \psi(r, r_2, \ldots) \psi(r', r_2, \ldots) \, dr_2 \ldots dr_A; \tag{2.12}$$

so the diagonal elements

$$\rho(r, r' = r) \equiv \rho(r) \tag{2.13}$$

give the density distribution.

By definition the momentum distribution is:

$$n(q) = \rho(q, q')|_{q'=q} \equiv \lim_{q' \to q} \int \psi(q, q_2, \ldots) \psi(q', q_2, \ldots) \frac{dq_2}{(2\pi)^3} \cdots \frac{dq_A}{(2\pi)^3}. \tag{2.14}$$

An explicit form of $n(q)$ expressed in terms of the one-body density matrix in coordinate space is as follows:

$$n(q) = \int \exp[iq \cdot (r - r')] \rho(r, r') \, dr \, dr'. \tag{2.15}$$

The Wigner distribution function $W(r, q)$ is also defined in terms of the one-body density matrix

$$W(r, q) = \int \rho(r + \tfrac{1}{2}\eta, r - \tfrac{1}{2}\eta) e^{iq \cdot \eta} \, d\eta, \tag{2.16}$$

and can be related to the density and momentum distributions:

$$\rho(r) = \int W(r, q) \frac{dq}{(2\pi)^3}, \tag{2.17}$$

$$n(q) = \int W(r, q) \, dr. \tag{2.18}$$

The relation of $\rho(r)$ and $n(q)$ to the Green function can be seen by using the relation between the one-particle Green function and the one-body density matrix:

$$\rho(r, r') = -i \lim_{t' \to t} G(r, t; r', t'). \tag{2.19}$$

The relation of $\rho(r, r')$ to $\rho(r)$ and $n(q)$ was already given by eqns (2.13) and (2.15).

2.2. Local density formalism

The local density distribution $\rho(r)$, one of the two most important quantities discussed in this book, has been well studied experimentally over a wide range of nuclei. This interest in $\rho(r)$ is related to the basic bulk nuclear characteristics such as the shape and sizes of nuclei, their binding energies, and other quantities connected with $\rho(r)$. Besides this the density distribution is an important object for experimental and theoretical investigations since it plays the role of a fundamental variable in nuclear theory. Theories are often formulated in terms of $\rho(r)$ and so other observables encountered are expressed in these terms as well. The pioneering approach in this direction was the Thomas–Fermi model (Thomas 1927; Fermi 1928).

Here we give a short description of this model using the concepts of quantum statistics.

Let us consider a fermion system of particles divided into regions large enough to be still considered as homogeneous, and containing enough particles for statistical considerations to be used. In a particular region characterized by a radius vector r the average kinetic energy per particle is expressed by the Fermi-momentum $k_F(r)$ (Thouless 1972):

$$\bar{E}_{\text{kin}} = \frac{3}{5} \frac{\hbar^2 k_F^2(r)}{2m} = \frac{3\hbar^2}{10m} \left(\frac{3\pi^2}{s}\right)^{\frac{2}{3}} \rho^{\frac{2}{3}}(r). \tag{2.20}$$

In eqn (2.20) we used the Fermi-gas relation between k_F and ρ: $k_F = (3\pi^2 \rho/s)^{\frac{1}{3}}$, where s is equal to 1 or 2 depending on whether the case in question is an electron Fermi-gas or a system of nucleons respectively.

The total kinetic energy of the whole system can be obtained using the expression

$$\int \bar{E}_{\text{kin}}(r)\rho(r) \, dr = \frac{3\hbar^2}{10m} \left(\frac{3\pi^2}{s}\right)^{\frac{2}{3}} \int \rho^{\frac{5}{3}}(r) \, dr \equiv \int \tau_{\text{TF}}(r) \, dr. \tag{2.21}$$

If $V(r)$ is the potential energy of the interaction of one particle with the other particles and eventually with an external field

$$V(r) = V_{\text{ext}}(r) + V_{\text{int}}(r), \tag{2.22}$$

then the total energy of the system is equal to

$$E = \int d\mathbf{r}\{\tau_{\text{TF}}(\mathbf{r}) + V_{\text{ext}}(\mathbf{r})\rho(\mathbf{r}) + V_{\text{int}}(\mathbf{r})\rho(\mathbf{r})\}. \tag{2.23}$$

This expression is correct if the potential energy is a slowly varying function of $\mathbf{r}$.

In the case of a many-electron atom the energy becomes a functional of the electron local density distribution

$$E = \int d\mathbf{r}\left\{\tau_{\text{TF}}(\mathbf{r}) - \frac{Ze^2}{r}\rho(\mathbf{r}) + \tfrac{1}{2}e^2\rho(\mathbf{r})\int\frac{\rho(\mathbf{r}')\,d\mathbf{r}'}{|\mathbf{r}-\mathbf{r}'|}\right\}, \tag{2.24}$$

where the second term in the integrand is the Coulomb potential-energy density of the interacting electrons with the electric charge of the nucleus (Ze). The third term reflects the interaction between electrons.

The condition for the minimum of the total energy (2.24) leads to the Euler–Lagrange equation:

$$\delta E/\delta\rho = \lambda, \tag{2.25}$$

with

$$\int \rho(\mathbf{r})\,d\mathbf{r} = Z. \tag{2.26}$$

Eqn (2.25) determines the equilibrium density ρ of the system and the electron-separation energy λ. The parameter λ plays the role of a Lagrange multiplier in the variational procedure subject to eqn (2.26).

The successful application of the Thomas–Fermi model to the atomic system is due to the fact that the one-particle potential can be easily introduced by a Coulomb electrostatic equation (the Poisson equation), which relates the Coulomb potential to the electron-charge distribution.

The lack of an analogous equation for the one-particle potential in the case of nuclear systems is the main difficulty in the application of the Thomas–Fermi method to nuclei. Nevertheless, many attempts have been undertaken to develop Thomas–Fermi-type approaches applicable to nuclear systems.

Here we briefly discuss the Thomas–Fermi approach of Bethe (1968) in which he applied the results of nuclear-matter theory to finite nuclear systems.

The total energy of the system is of the form:

$$E = \int W(\rho)\rho(\mathbf{r})\,d\mathbf{r} + \frac{1}{2}\int u(\mathbf{r},\mathbf{r}')[\rho(\mathbf{r}) - \rho(\mathbf{r}')]^2\,d\mathbf{r}\,d\mathbf{r}', \tag{2.27}$$

where $W(\rho)$ is the energy per nucleon in nuclear matter, including the

Thomas–Fermi kinetic-energy expression

$$\tau_{\mathrm{TF}}(r) = \frac{3\hbar^2}{10m} k_{\mathrm{F}}^2(r)\rho(r), \tag{2.28}$$

and the function $u(r, r')$ describes the effective interaction between nucleons at the points r and r'. The Euler–Lagrange variational equation related to the expression (2.27) has been analyzed in detail. It turns out that the equation of motion does not have a continuous solution (Siemens 1970) for the density $\rho(r)$. As pointed out by Siemens (1970) the reasons for this lie in the replacement of the attractive exchange interaction by the effective local one.

In parallel with the efforts made to develop the Thomas–Fermi nuclear method, phenomenological approaches have been suggested for the description of nuclear bulk properties like binding and deformation energies, density distributions and radii, and high-lying collective excitations. In the energy-density formalism EDF (Brueckner *et al.* 1968, 1969*a*, 1969*b*; Lombard 1973) the total energy

$$E = \int \varepsilon[\rho(r)] \, dr \tag{2.29}$$

is written in terms of the energy-density functional $\varepsilon[\rho]$:

$$\begin{aligned}\varepsilon[\rho] = &\, 0.3(\hbar^2/M)(\tfrac{3}{2}\pi^2)^{\frac{2}{3}} \tfrac{1}{2}[(1+\alpha)^{\frac{5}{3}} + (1-\alpha)^{\frac{5}{3}}]\rho^{\frac{5}{3}} \\ & + \rho V(\rho, \alpha) + \tfrac{1}{2}e\rho_p \Phi_{\mathrm{C}}(r) - 0.7386 e^2 \rho_p^{\frac{4}{3}} \\ & + \frac{\hbar^2}{8m}\eta(\nabla\rho)^2 + c\frac{(\nabla\rho)^2}{\rho},\end{aligned} \tag{2.30}$$

where

$$\alpha = \frac{\rho_n - \rho_p}{\rho_n + \rho_p}, \qquad \rho = \rho_n + \rho_p, \tag{2.31}$$

$$V(\rho, \alpha) = b_1\rho + b_2\rho^{\frac{4}{3}} + b_3\rho^{\frac{5}{3}} + \alpha^2(b_4\rho + b_5\rho^{\frac{4}{3}} + b_6\rho^{\frac{5}{3}}), \tag{2.32}$$

and $\Phi_{\mathrm{C}}(r)$ is the Coulomb potential connected with the charge distribution ρ_p.

The first term in eqn (2.30) is the Thomas–Fermi kinetic energy, and the second is the potential energy taken from nuclear-matter theory. The third and fourth terms are the Coulomb direct and exchange terms respectively. The gradient corrections to the potential and kinetic energy are given by the last two terms. The parameters b_1, b_2, b_3 in (2.32) refer to symmetric ($N = Z$) nuclear matter, and b_4, b_5, b_6 to non-symmetric ($N \neq Z$) nuclear matter (Brueckner *et al.* 1969*b*).

Many analyses of the usefulness of the energy-density formalism for the study of nuclear ground-state properties have been made (Lombard 1973), together with detailed studies of the role of particular terms. On the whole this phenomenological approach gives good overall results for the bulk characteristics of nuclei. An unsatisfactory feature is the presence of adjustable parameters in the theory, and some of these (η and c) are connected with terms of different physical meaning: the term $\hbar^2\eta(\nabla\rho)^2/8m$ is related to the potential energy, while $c(\nabla\rho)^2/\rho$ (the so-called Weizsäcker correction) is related to the kinetic energy.

In the extended Thomas–Fermi (ETF) approach (Bohigas *et al.* 1976; Kirzhnitz 1967; Guet and Brack 1980; Brack *et al.* 1985; Ring and Schuck 1980; Bartel *et al.* 1985) the semiclassical expression for the kinetic-energy density functional is used together with the Thomas–Fermi kinetic-energy term:

$$\tau_{\text{ETF}}[\rho] = \tau_{\text{TF}}[\rho] + \tau_2[\rho] + \tau_4[\rho], \qquad (2.33)$$

where τ_2 and τ_4 contain the density ρ and its gradient and are proportional to $\hbar^2$ and $\hbar^4$ respectively. The term proportional to $\hbar^6$ has been considered (Kirzhnitz 1967) and shown to lead to divergences (Guet and Brack 1980).

In ETF approach the energy functional is given by

$$E_{\text{ETF}} = \int d\mathbf{r}[V(\mathbf{r})\rho(\mathbf{r}) + (\hbar^2/2m)(\tau_{\text{TF}}[\rho] + \tau_2[\rho] + \tau_4[\rho])], \qquad (2.34)$$

where $V(\mathbf{r})$ is a given local (Hartree–Fock) potential connected to Skyrme-type effective interactions.

Calculations in the framework of the ETF approach give a quantitative agreement with the results of average Hartree–Fock calculations, as used in the Strutinsky method for nuclear densities, and for binding and deformation energies. After perturbative inclusion of the shell effects by a single Hartree–Fock iteration, the exact total Hartree–Fock energies for heavy nuclei are reproduced quite satisfactorily. An interesting application of this method is the calculation of fission barriers, and it is found that the most reasonable results are obtained using the SkM* set (Krivine *et al.* 1980) of Skyrme-type parameters.

The ETF method needs further development of its mathematical formulation and applications. Special attention deserves to be paid to the asymptotic behaviour of the density distribution (in the work of Brack *et al.* 1985 the asymptotic form is $\sim -c/r^6$) and to the applicability of a Ritz-variational procedure, which at present leads to an over-estimation of the binding energies of some nuclei.

The problem of the formulation of nuclear theory only in terms of local density $\rho(\mathbf{r})$ (as described above) is difficult because of the A-

representability limitation (Ghosh and Deb 1982). In the approaches mentioned in this section it is not obvious that the local densities ρ satisfy the necessary conditions for it to be derivable from a properly antisymmetrized wavefunction.

The theorem for the existence of a ground-state energy functional of ρ which, in principle, justifies the local density formulation of the theory is given in the next section.

2.3. The Hohenberg–Kohn theorem and the general functional relation between the nucleon momentum and density distributions

There are several reasons which underline the importance of the formulation of the local-density variant of the theory. It is crucial that ρ is a much simpler 3-dimensional object irrespective of the number of particles. This is the great simplification of the theory compared with the many-particle wavefunction description. In addition the density distribution as an observable quantity can be measured experimentally and thus the calculations and different approximations in the framework of the local density theory can be tested.

An important role in the development of the density-functional theory is played by the Hohenberg–Kohn (HK) theorem (Hohenberg and Kohn 1964; Kohn and Sham 1965). This gives a solid mathematical basis and formal justification for the use of the density as a fundamental quantity.

Hohenberg and Kohn (1964) prove a theorem establishing the existence of a functional $E[\rho]$ representing the total energy of the fermion system as a universal (perhaps unique) functional of the local density. The theorem establishes that the many-particle wavefunction and hence all ground-state properties of the system are unique functionals of the density, and that the energy functional has a minimum value for the exact density.

The density ρ and the binding energy E can be variationally determined by minimizing the energy functional $E[\rho]$ with respect to the density, taking into account the condition

$$\int \rho(r)\,dr = A. \tag{2.35}$$

The stationary condition has the general form

$$\delta\left\{E[\rho] - \mu \int \rho(r)\,dr\right\} = 0, \tag{2.36}$$

or equivalently

$$\delta E/\delta\rho - \mu = 0, \tag{2.37}$$

where μ is a Lagrange multiplier corresponding to the constraint (2.35).

The construction problem of the energy functional $E[\rho]$ is not treated by the HK theorem. This theorem makes an abstract statement but does not give a way of finding an explicit form of the functional. The building of particular forms of the functionals requires physical assumptions and approximations. Several examples were given in the previous section.

The energy-density theory has been elaborated and precisely formulated and applied to a wide variety of problems in atomic, molecular, solid state, and nuclear physics. Many works devoted to different aspects and variants of the HK theorem and applications of energy-density functional theory are reviewed in the papers of Ghosh and Deb (1982) and Gupta and Rajagopal (1982).

Now we arrive at (Antonov and Petkov 1986) some consequences of the HK theorem relevant to the basis quantities discussed in this book, namely the nucleon density and momentum distributions. We point out that there must be a functional relation between the momentum and density distributions.

Indeed, the many-particle wavefunction ψ is a functional of $\rho(r)$ according to the HK theorem. It is

$$\psi \to \psi(r_1, r_2, \ldots r_A; [\rho]) \equiv \psi[\rho]. \quad (2.38)$$

We know that the one-body density matrix $\rho(r, r')$ is determined by the many-particle wavefunction:

$$\rho(r, r') = \int \psi(r, r_2, \ldots; [\rho]) \psi(r', r_2, \ldots; [\rho]) \, dr_2 \ldots dr_A \equiv \rho(r, r'; [\rho]). \quad (2.39)$$

Obviously the one-body density matrix $\rho(r, r')$ becomes a functional of the density $\rho(r)$. In this way the HK theorem establishes the principal functional relation between the non-diagonal and diagonal elements of the one-body density matrix.

Now it immediately follows that the momentum distribution $n(k)$, which is the diagonal element of the Fourier transform of $\rho(r, r'; [\rho])$ is also the functional of $\rho(r)$. Thus we come to the following important general relation:

$$n(k) = \int \exp i k \cdot (r - r') \rho(r, r'; [\rho]) \, dr \, dr' \equiv n(k; [\rho]). \quad (2.40)$$

The relation (2.40) is also an abstract statement analogous to the HK theorem, so there is no constructive way to build the functional $n(k; [\rho])$. Its eventual applications requires additional physical assumptions and approximations.

An approximate explicit relation between $n(\mathbf{k})$ and $\rho(\mathbf{r})$, which can be considered as an illustration of the statement (2.40) is given by the coherent density fluctuation model (CDFM) (Antonov et al. 1979, 1980, 1986b) discussed in detail in Section 8.3 of this book.

The existence of the relation (2.40) allows us to formulate the variational principle in terms of both quantities $n(\mathbf{k})$ and $\rho(\mathbf{r})$. Indeed, as the energy of the system is determined variationally from the functional $E[\rho]$ and as $\rho(\mathbf{r})$ and $n(\mathbf{k})$ are related, it follows quite generally that E is functionally dependent on ρ and n

$$E[\rho] \to E[\rho, n]. \tag{2.41}$$

Furthermore, it can be shown that $n(\mathbf{k})$ is a unique functional of ρ. Arguments for this are provided by the HK theorem. Moreover this theorem can be proved in terms of $n(\mathbf{k})$ as well. For this purpose, let us consider a system of fermions with Hamiltonian,

$$H = T + U + V, \tag{2.42}$$

where

$$T = \int \frac{d\mathbf{p}}{(2\pi)^3} \frac{\hat{p}^2}{2m} a^+(\mathbf{p})a(\mathbf{p}) \tag{2.43}$$

is the kinetic-energy operator in second-quantized form, and

$$U = \frac{1}{2} \int \frac{d\mathbf{p}}{(2\pi)^3} \frac{d\mathbf{p}'}{(2\pi)^3} \frac{d\mathbf{p}''}{(2\pi)^3} \tilde{V}_{NN}(\mathbf{p}'' - \mathbf{p}') a^+(\mathbf{p}'') a^+(\mathbf{p} + \mathbf{p}' - \mathbf{p}'')$$
$$\times a(\mathbf{p}')a(\mathbf{p}) \tag{2.44}$$

is the interaction operator between particles,

$$\tilde{V}_{NN}(\mathbf{p}'' - \mathbf{p}') = \int d\mathbf{r}\, V_{NN}(\mathbf{r}) \exp[-i(\mathbf{p}'' - \mathbf{p}')\mathbf{r}], \tag{2.44a}$$

and

$$V = \int \frac{d\mathbf{p}}{(2\pi)^3} \frac{d\mathbf{p}'}{(2\pi)^3} \tilde{V}(\mathbf{p} - \mathbf{p}') a^+(\mathbf{p})a(\mathbf{p}') \tag{2.45}$$

is an operator corresponding to the external field $\vartheta(\mathbf{r})$ and

$$\tilde{V}(\mathbf{p} - \mathbf{p}') = \int d\mathbf{r}\, \vartheta(\mathbf{r}) \exp[-i(\mathbf{p} - \mathbf{p}')\mathbf{r}]. \tag{2.46}$$

Following the original proof of the HK theorem one can state that the momentum distribution in the ground state ψ:

$$n(\mathbf{k}) = \langle \psi | a^+(\mathbf{k})a(\mathbf{k}) | \psi \rangle \tag{2.47}$$

is a unique functional of $\vartheta(r)$ and, conversely, $\vartheta(r)$ is unique functional of $n(k)$. Since in turn $\vartheta(r)$ fixes H one can see that the ground state is a unique functional of $n(k)$. In the same way as in the HK theorem it can be shown that the energy $E[n]$ has its minimum for the correct $n(k)$ for which the condition holds:

$$\int n(k) \frac{dk}{(2\pi)^3} = A. \tag{2.48}$$

Now since $\rho(r)$ and $n(k)$ are uniquely related to $\vartheta(r)$ it follows that the functional relation between them will also be unique.

This consideration makes it possible to formulate the variational principle in terms of the variables $\rho(r)$ and $n(k)$, keeping in mind the principal relation between them as stated above. Such a formulation can prove useful in practical applications when the functional relation (2.40) is guessed on the basis of experimental facts or in the context of reasonable phenomenology.

The given formulation in which the two basic characteristics of the system $\rho(r)$ and $n(k)$ enter in the same way is interesting in the pure theoretical aspects. For example, since $n(k)$ is sensitive to the nucleon-nucleon correlations at small distances, the energy functional $E[\rho, n]$ must be defined in agreement with this restriction, properly defining the forces between the particles. We note that the formulation of the theory in terms of $\rho(r)$ (for example, Thomas–Fermi-like models) or of $n(k)$ (Landau–Fermi-liquid theory) can now be considered, in principle, as contained in this more general 'mixed' (ρ and n) formulation of the theory.

The variational principle can be formulated as follows:

In order to minimize E for independent variations of ρ and n we must consider the following functional:

$$\tilde{E}[n, \rho] = E[n, \rho] - E_F \int \rho \, dr - \int g(k)[n(k) - n(k; [\rho])] \frac{dk}{(2\pi)^3}, \tag{2.49}$$

where the second and third terms reflect the constraints:

$$\int \rho(r) \, dr = A, \tag{2.50}$$

$$n(k) = n(k; [\rho]). \tag{2.51}$$

The variational principle now leads to the Euler–Lagrange equations of the form:

$$\frac{\delta E}{\delta \rho} + \int \frac{\mathrm{d}\boldsymbol{k}}{(2\pi)^3} g(\boldsymbol{k}) \frac{\delta n(\boldsymbol{k};[\rho])}{\delta \rho} = E_F, \qquad (2.52)$$

$$\delta E/\delta n = g(\boldsymbol{k})/(2\pi)^3, \qquad (2.53)$$

$$\int \rho \, \mathrm{d}\boldsymbol{r} = A, \qquad (2.54)$$

$$n(\boldsymbol{k}) = n[\boldsymbol{k};[\rho]]. \qquad (2.55)$$

The solution of eqns (2.52)–(2.55) gives three functions $\rho(\boldsymbol{r})$, $n(\boldsymbol{k})$, $g(\boldsymbol{k})$, and the Lagrange parameter E_F which determines the particle-separation energy. Since eqn (2.53) determines the function $g(\boldsymbol{k})$, eqn (2.52) can be rewritten in the form:

$$\frac{\delta E}{\delta \rho} + \int \frac{\delta E[n, \rho]}{\delta n} \frac{\delta n(\boldsymbol{k};[\rho])}{\delta \rho} \mathrm{d}\boldsymbol{k} = E_F. \qquad (2.56)$$

As an illustration let us consider the simplest case when only the kinetic energy is taken as a functional of n:

$$T[n] = \int \frac{\hbar^2 k^2}{2m} n(\boldsymbol{k}) \frac{\mathrm{d}\boldsymbol{k}}{(2\pi)^3}. \qquad (2.57)$$

In this case

$$E[\rho, n] = T[n] + V[\rho], \qquad (2.58)$$

with $V[\rho]$ the total potential energy of the system. Then eqn (2.56) can be written in a simple tractable form:

$$\frac{\delta}{\delta \rho} t[\rho] + \frac{\mathrm{d}}{\mathrm{d}\rho}(\rho \vartheta) = E_F, \qquad (2.59)$$

where

$$t[\rho] = \int \frac{\hbar^2 k^2}{2m} n(\boldsymbol{k};[\rho]) \frac{\mathrm{d}\boldsymbol{k}}{(2\pi)^3}, \qquad (2.60)$$

$$\frac{\mathrm{d}}{\mathrm{d}\rho}(\rho \vartheta(\rho)) = \frac{\delta V}{\delta \rho}, \qquad (2.61)$$

where $V[\rho]$ can be represented in the form

$$V[\rho] = \int \mathrm{d}\boldsymbol{r} \, \rho(\boldsymbol{r}) \vartheta(\rho). \qquad (2.62)$$

The quantities $\delta t[\rho]/\delta \rho$ and $\mathrm{d}(\rho \vartheta)/\mathrm{d}\rho$ are the one-particle kinetic and potential energies respectively.

Let us consider the case when

$$n(k;[\rho]) = \int dr\, n(\rho, k), \qquad (2.63)$$

where

$$n(\rho, k) = 4\theta[k_F(\rho(r)) - |k|]. \qquad (2.64)$$

Then from (2.60) and (2.63) we obtain

$$\frac{\delta}{\delta\rho} t[\rho] = \frac{d}{d\rho}[\rho T_{TF}(k_F(\rho))], \qquad (2.65)$$

where

$$T_{TF}(k_F(\rho)) = \frac{3\hbar^2}{10m} k_F^2(\rho) \qquad (2.66)$$

is the Thomas–Fermi kinetic energy. Then eqn (2.59) can be rewritten in the form:

$$\frac{d}{d\rho}[\rho T_{TF}(k_F(\rho))] + \frac{\delta V}{\delta\rho} = E_F. \qquad (2.67)$$

So one can see that the Thomas–Fermi model relations are obtained as a particular case of the approach assuming the concrete form of the functional dependence (2.63)–(2.64). Note that this relation has been introduced by Hüfner and Nemes (1981) as a local Fermi-gas approximation.

Now it is easy to obtain the non-interacting Fermi-gas model ($V = 0$, $\rho = \rho_0 =$ constant) relations. From (2.63), (2.64), (2.66), and (2.67) we obtain:

$$n(k;[\rho]) = 4\frac{A}{\rho_0}\theta(k_F - k), \qquad (2.68)$$

$$T_F = 3\hbar^2 k_F^2/10m, \qquad E_F = \hbar^2 k_F^2/2m, \qquad k_F = (\tfrac{3}{2}\pi^2\rho_0)^{\tfrac{1}{3}}. \qquad (2.69)$$

A more complicated density-functional form of $n(k;[\rho])$ for the finite nuclei with monotonic decreasing density distribution $\rho(r)$ has been suggested by Antonov et al. (1979, 1980) and will be considered in detail in Section 8.3 of this book:

$$n(k) = \left(\frac{4\pi}{3}\right)^2 \frac{4}{A}\left[6\int_0^{\alpha/k} \rho(x)x^5\, dx - \left(\frac{\alpha}{k}\right)^6 \rho\left(\frac{\alpha}{k}\right)\right] \equiv n(k;[\rho]),$$

$$\int n(k)\frac{dk}{(2\pi)^3} = A, \qquad \alpha = (\tfrac{9}{8}\pi A)^{\tfrac{1}{3}}.$$

Note that the term linear in ρ in (2.70) ensures the uniqueness of the functional $n(k;[\rho])$. The functional form $n(k;[\rho])$ in the more general case of non-monotonic density distributions was proposed by Antonov et al. (1986b).

Following the scheme for obtaining the variational equation (2.59) we get for $n(k)$ given in (2.70):

$$\frac{\alpha^2\hbar^2}{10m}\frac{1}{r^2} + \frac{d}{d\rho}(\rho\vartheta) - \frac{\hbar^2}{4m}\eta\,\Delta\rho = E_F. \qquad (2.71)$$

Here the potential energy $V[\rho]$ has taken the form (Lombard 1973):

$$V[\rho] = \int d\mathbf{r}\left[\rho(r)\vartheta(\rho) + \frac{\hbar^2}{8m}\eta(\nabla\rho)^2\right]. \qquad (2.72)$$

In general, eqn (2.71) must be solved with respect to ρ, and then the relation (2.54) must be used in order to obtain the Lagrange parameter E_F. This solution $\rho(r, E_F) \equiv \rho_0(r)$ determines $n_0(k)$ from (2.70) and the binding energy of the system $E[n_0, \rho_0]$.

A possible estimate of E_F in (2.71) following the practical method of March (1979) can be obtained by multiplying (2.71) by $\rho(r)$ and integrating over r. Then we get

$$\bar{E}_F = \frac{1}{A}\int \rho(r)\left[\frac{\alpha^2\hbar^2}{10m}\frac{1}{r^2} + \frac{d}{d\rho}(\rho\vartheta) - \frac{\hbar^2}{4m}\eta\,\Delta\rho\right]dr. \qquad (2.73)$$

The evaluation of $\bar{E}_F$ needs the specification of the potential ϑ. In order to see how reasonable is the functional (2.70) one can estimate the average separation energy $\bar{E}_F$ (2.73) using the potential ϑ in the form given in the papers of Brueckner (1969a) and Lombard (1973) and the parameter $\eta = 12.0$. The nuclear density $\rho(r)$ can be taken in a symmetrized Fermi-type form $\rho(r; R, b)$ with parameters R and b determined from the electron-scattering experimental data (Burov et al. 1974). In Table 2.1 the results of the calculations of $\bar{E}_F$ for ^{16}O and ^{40}Ca are listed and compared with the experimental separation energies.

TABLE 2.1. Proton and neutron separation energies in ^{16}O and ^{40}Ca.

	Separation energy (MeV)			
	Proton		Neutron	
Nucleus	From (2.73)	Experiment	From (2.73)	Experiment
^{16}O	11.70	12.127	12.885	15.669
^{40}Ca	14.42	8.330	16.92	15.636

It can be seen that though the procedure is not self-consistent it leads to a reasonable qualitative estimate of $\bar{E}_F$ especially for the separation energy of a proton in ^{16}O and of a neutron in ^{40}Ca. The realization of the full computational procedure (i.e. self-consistent determination of $\rho(r)$ on the base of (2.71) and consequently of the momentum distribution $n(k)$ (2.70)) needs the specification of the forces at small r because of the $\sim 1/r^2$ term in eqn (2.71) originating from the particular form of $n(k)$ (2.70) or the kinetic energy.

In conclusion we note that the development of this approach is related to the refinement of the functional-density dependence of the nucleon momentum distribution and the nuclear forces, especially of their short-range behaviour. This programme in practice needs, as well as the usual data necessary for a determination of the nucleon–nucleon forces (phase-shift analysis, binding energies etc.), one additional nuclear characteristic, namely the nucleon momentum distribution. The results for the density distribution $\rho(r)$, binding energies, and other bulk properties would then be consistent both with their experimental values and with the momentum distribution of the system.

3

GENERAL CONSIDERATIONS RELATED TO NUCLEON MOMENTUM DISTRIBUTIONS

In this chapter the qualitative features of the nucleon momentum distributions are discussed in their relation to the main characteristics of nucleon-nucleon forces as well as to the possibility of extracting the nucleon momentum distribution from the experimental data. In Section 3.1 the connection of the high-momentum behaviour of the momentum distribution with the short-range nucleon-nucleon correlations will be demonstrated, based on the method of Brueckner *et al.* (1955). Several reactions are considered in Section 3.2 illustrating the principal relations between the nucleon momentum distribution and the corresponding cross-sections. Interpretation of experimental data shows model dependence.

The dependence of the nucleon momentum distribution on nucleon-nucleon forces at small distances and its asymptotic behaviour at large momenta are discussed in the framework of a schematic solvable one-dimensional model in Section 3.3.

3.1. Qualitative arguments concerning the nucleon momentum distribution

Systematic investigation of momentum distributions extended the scope of nuclear ground-state theory. Some decades ago nuclear-matter theories were usually limited to the calculation of a small number of quantities, for example the equilibrium density, the binding energy per nucleon, and the symmetry energy.

The experimental situation in recent years concerning the interaction of particles at high energies, in particular the (p, 2p), (e, e'p) reactions, the nuclear photo-effect, meson absorption by nuclei, inclusive proton production in high-energy proton-nucleus collisions, and even some phenomena at low energies such as giant multipole resonances, makes it possible to study additional quantities contained in the theory. Such quantities specifically related to the above mentioned processes are the nucleon momentum distribution $n(k)$ and the pair-correlation function

NUCLEON MOMENTUM DISTRIBUTIONS

$\rho(r_1, r_2; r_1, r_2)$ (1.30). Here we shall consider only the main subject of this book, the nucleon momentum distribution. The qualitative knowledge of this distribution is very important for revealing the proper mechanism of nuclear reactions and for their complete description (Gottfried 1963).

In order to understand the main features of the nucleon momentum distribution we start by considering some simple model systems:

(1) In the case of the free Fermi-gas the momentum distribution is a step function as illustrated in Fig. 3.1.

(2) A dilute hard-sphere Fermi-gas has a momentum distribution with a small but rather long high-momentum tail and a discontinuity at k_F (Czyż and Gottfried 1961) (Fig. 3.2).

(3) For non-interacting fermions in a shell-model-type potential well, $n(k)$ is shown in Fig. 3.3. In this case $n(k)$ no longer has a discontinuity and the high-momentum tail is now very short (in the case when $A^{\frac{1}{3}} \gg 1$).

For a real nucleus the detailed form of $n(k)$ at $k \leq k_F$ is determined from the long-range attractive potential, while the behaviour of $n(k)$ for k sufficiently larger than k_F should be dominated by the effects of the hard core and the strong attraction just outside it. As the long-tail behaviour of $n(k)$ is connected to this feature of the nucleon–nucleon force one can hope that it should not differ greatly between the finite nucleus and nuclear matter.

The nuclear shell-model approaches give reasonable results for $n(k)$ in the region of $k \leq k_F$ where the momentum distribution is insensitive to correlation effects. The presence of the short-range correlations produces effects not in accord with shell-model predictions at $k > k_F$. Qualitatively this could be understood by noting that the shell-model potential is rather

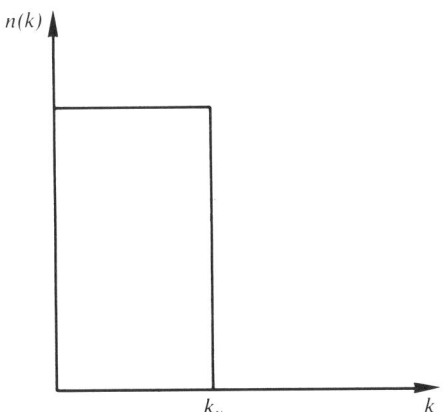

FIG. 3.1. Momentum distribution for a free infinitely-extended Fermi-gas.

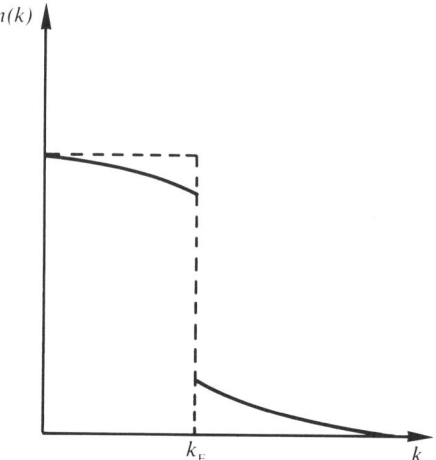

Fig. 3.2. Momentum distribution for a dilute hard-sphere Fermi-gas.

smooth, so the relevant shell-model wavefunctions do not contain high-momentum Fourier components. The shell-model does not contain the possibility of two particles coming close to each other, which is a necessary condition for their acquisition of rather high momenta. This makes it necessary to develop theoretical approaches which take into account the short-range correlations due to the character of the nucleon–nucleon forces at small distances. This should be done so as to include as

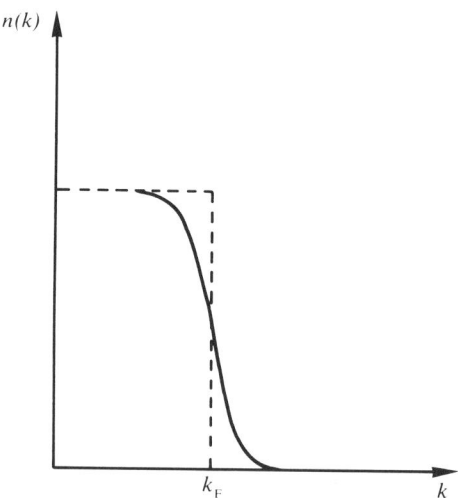

Fig. 3.3. A sketch of $n(k)$ for a shell-model wavefunction ($A \gg 1$).

a particular case the usual shell-model approach. An attempt in this direction is made by Brueckner *et al.* (1955). The ground-state function $\psi(\{r_i\})$ ($i = 1, 2, \ldots A$) is constructed on a shell-model function basis $\Phi(\{r_i\})$ by the transformation function, or 'model operator' F, i.e.

$$\psi(\{r_i\}) = F\Phi(\{r_i\}). \tag{3.1}$$

Since Φ is a weakly correlated wavefunction, F has the effect of introducing correlations into it. The explicit form of the transformation is given by the following set of coupled equations:

$$F = 1 + \frac{1}{e}\sum_{ij} I_{ij} F_{ij}, \tag{3.2}$$

$$F_{ij} = 1 + \frac{1}{e}\sum_{lm \neq ij} I_{lm} F_{lm}, \tag{3.3}$$

$$e = E_0 - \sum_i T_i - V_C. \tag{3.4}$$

The energy eigenvalue is determined by:

$$(E_0 - \Sigma T_i - V_C)\Phi = 0. \tag{3.5}$$

The two-body scattering operators t_{ij} are defined by:

$$t_{ij} = V_{ij} + V_{ij}(1/e)t_{ij}, \tag{3.6}$$

where V_{ij} is the potential between nucleons i and j. They determine the operators I_{ij} as the parts of t_{ij} which are non-diagonal with respect to the nuclear states and the uniform potential V_C:

$$V_C = \tfrac{1}{2}\sum_{ij} t_{Cij}, \tag{3.7}$$

where t_{Cij} is the diagonal part of t_{ij}.

The non-diagonal operators I_{ij} cause transitions from the uncorrelated independent-particle states $\Phi(\{r_i\})$; the effect is closely analogous to an inelastic scattering of particles, a pair at a time, out of the Fermi-gas to excited states. Consequently the departures of the wavefunction ψ from Φ are very closely related to the details of the inelastic scattering of nucleons by nucleons and thus to the strength and range of the two-body potentials.

Although this consideration is brief and not complete it reveals the direct relation of the nucleon-nucleon correlations at short distances (the deviation of ψ from Φ) to the nucleon-nucleon forces in terms of which the operator F is expressed.

3.2. Determination of the nucleon momentum distribution from experimental data

The principal relation between the nucleon momentum distribution and high-energy nuclear reactions can be illustrated by the example of the inclusive backward-particle production reaction. Frankel et al. (1976) investigate cross-sections of a 180° production of high-energy protons, deuterons, and tritons as a result of proton collisions (with energies 600 MeV and 800 MeV) on Be, C, Cu, Ta, Ag, and Pt. The production of particles in the region kinematically forbidden in the elementary (free) particle events (at the same incident energy) was interpreted assuming a single-particle mechanism for the reaction (see Fig. 3.4) (Amado and Woloshyn 1976a) in which the incident proton with momentum p (laboratory system) strikes the target nucleon with momentum k. The final state consists of the observable particle with momentum q and other products which are not detected.

The cross-section of the reaction (Amado and Woloshyn 1976a) is given by:

$$\frac{d\sigma}{d^3q} = \frac{M^4}{pE(q)} \frac{1}{2(2\pi)^5} \int \frac{dk}{E(k)E(p+k-q)} [n_p(k)\Sigma|m_{pp}|^2 + n_n(k)\Sigma|m_{pn}|^2]\delta[E(p) + M - E(q) - E(p+k-q) - \bar{\varepsilon}],$$

(3.8)

where $n_p(k)$ and $n_n(k)$ are the proton and neutron momentum distributions respectively, $\Sigma|m|^2$ is the square of the nucleon–nucleon amplitude summed over spin, $\bar{\varepsilon}$ is the average nucleon interaction energy and $E(p) = (p^2 + M^2)^{\frac{1}{2}}$, M being the nucleon mass. The main feature of (3.8) we want to emphasize is the presence of the momentum distributions $n_{p(n)}(k)$ in the expression for the cross-section. Taking into account the δ

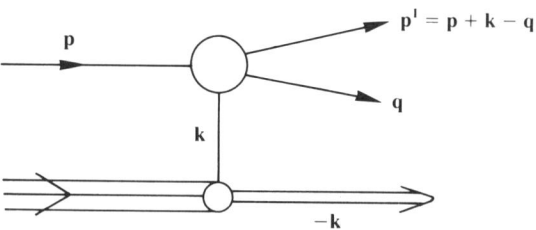

FIG. 3.4. Single-particle mechanism for inclusive proton production in proton-nucleus collisions (Amado and Woloshyn 1976a).

function we note that the integral over k is between

$$k_{\min} = |\mathbf{p} - \mathbf{q}| - (2M\omega + \omega^2)^{\frac{1}{2}},$$
and
$$k_{\max} = |\mathbf{p} - \mathbf{q}| + (2M\omega + \omega^2)^{\frac{1}{2}}, \quad (3.9)$$

where $k_{\min}$ in these processes lies in the high-momentum region: 700–1400 MeV/c ($\cong$3.5–7.0 fm^{-1}), which is larger than the characteristic Fermi-momentum $k_F \sim \rho^{\frac{1}{3}} \cong 1.5$ fm^{-1}.

The experimental data for these reactions have been analysed by Amado and Woloshyn (1976a) who suggest the following two equivalent phenomenological forms for the momentum distributions:

$$n_s(k) = N_c k \gamma_s / \sinh(\gamma_s k), \quad (3.10a)$$

$$n_c(k) = N_c / \cosh^2(\gamma_c k), \quad (3.10b)$$

where N_c is a normalization constant and γ_s and γ_c are scale parameters. The correct behaviour of these distributions at large k ($\sim \exp(-\gamma_s k)$ and $\sim \exp(-2\gamma_c k)$) is essential for a reasonable description of the experimental cross-section. The distributions (3.10) strongly deviate from the known Fermi-gas momentum distribution which fails to reproduce the experimental data. The momentum distributions (3.10a,b) will be discussed in more detail in Section 3.3.

The information we obtain for the momentum distribution obviously depends on the particular reaction mechanism used in the analysis of the experimental data.

For example if we use another mechanism (Weber and Miller 1977), according to which the incident particle interacts with $A - 1$ particles (Fig. 3.5), the suggested realistic nuclear high-momentum distribution also leads to a qualitative explanation of the data.

In the model suggested by Burov et al. (1977) the clusters of several nucleons ($N = 2, 3, 4, \ldots$) can be formed in a fairly small volume (the so-called 'fluctons'). Using this model the cross-sections of pion-inclusive

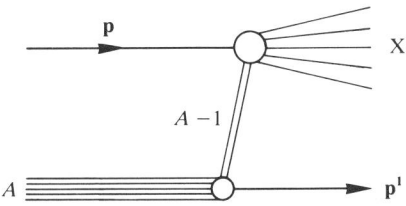

FIG. 3.5. Mechanism for inclusive proton production in proton-nucleus collisions (Weber and Miller 1977).

reactions in 8 GeV proton-nucleus collisions have been successfully analyzed.

A similar approach of 'correlated clusters' has been used by Fujita (1977) for the explanation of the proton-nucleus experiments of Frankel.

A hard-scattering mechanism was proposed by Hatch and Koonin (1979) to describe proton and pion spectra from high-energy (800 MeV/nucleon) nucleus–nucleus (C + C, Ne + NaF, C + Pb) collisions by means of the empirical nucleon momentum distribution (3.10a).

The problem of extracting information for $n(k)$ from the high-energy experiments becomes very complicated if one takes into account the final-state interaction. As was shown by Amado and Woloshyn (1977a), the large-momentum result cannot be simply interpreted in terms of a momentum distribution of the target nucleons. They conclude that the momentum distribution is an extremely difficult quantity to determine experimentally.

This problem is discussed by Frankel (1978) on the basis of the proposed 'quasi two-body scaling' approach (Frankel 1977). The cross-section in this method is related to the momentum distribution as follows:

$$\frac{d\sigma}{d^3q} = \frac{C(p, k_{min})G(k_{min})}{|p - q|}, \qquad (3.11)$$

where p and q are the momenta of the incident projectile and outgoing particles; p' is the momentum of the scattered projectile, and

$$|k_{min}| = |p - q| - |p'|, \qquad (3.12)$$

$$G(k_{min}) = \int_{k_{min}}^{\infty} n(k)k \, dk. \qquad (3.13)$$

The quantity $G(k_{min})/|p - q|$ is the 'probability' of obtaining a recoiling $A - 1$ nucleus of momentum k_{min}; and $C(p, k_{min})$ describes the p and k_{min} dependence of the cross-section for the inclusive process $p + A \to p + p' + (A - 1$ nucleus).

By critical consideration of the role of the final-state interaction (Amado and Woloshyn 1977a) it has been shown (Frankel 1978) that the momentum distribution $n(k)$ in (3.13) has to be interpreted as an effective one, or $n_{eff}(k)$, which is roughly proportional to $n(k)$. Thus the direct relation between the measured cross-section and the real nucleon momentum distribution in nuclei still remains open. This indicates the need of further systematic investigations of both the theoretical and experimental aspects. Reactions with different projectiles such as: protons (Frankel et al. 1978a,1978b; Cordell et al. 1981a; Roy et al. 1981; Komarov et al. 1978,1979; Avan et al. 1984); deep inelastic lepton scattering from nuclear targets (Bodek and Ritchie 1981; Bernheim et al.

1981; Nagorny et al. 1985); alpha-particles (Cordell et al. 1981b); deuterons (Azhgirey et al. 1977); observed particles (Boal and Woloshyn 1979; Landau 1978); as well as different types of reactions: $(\pi^+, \pi^+ p)$ and $(\pi^+, 2p)$ (Grashin and Shalamov 1979); $(\alpha, {}^3\text{He})$ and $({}^{16}\text{O}, {}^{15}\text{O})$ (Hüfner and Nemes 1981; Fujita and Hüfner 1980; Bertsch 1981; Hiller and Hüfner 1982; Araseki and Fujita 1985); (p, 2p) (Boal 1980). Charged-particle emission following muon capture in complex nuclei (Lifshitz and Singer 1978) are of special interest for the purpose of obtaining reliable information on the nucleon momentum distribution.

In the paper of Fujita (1986) are considered the kinematical conditions, in which use of the one-nucleon knockout-reaction mechanism using momentum distributions with high-momentum components generated by the short-range nucleon–nucleon correlations, can be justified.

3.3. A schematic solvable model

Since the behaviour of the nucleon momentum distribution at large k depends essentially on the character of the nucleon–nucleon forces at small distances (or nucleon–nucleon correlations effects) one can hope that schematic but exact-solvable models should give correctly the general features of the nucleon momentum distribution at large k.

Here we give briefly the model developed in the papers of Amado (1976), Amado and Woloshyn (1976b, 1977b), in which an N-particle, one-dimensional Hamiltonian with delta-forces is chosen as follows ($\hbar = 2m = 1$):

$$H = -\sum_{i=1}^{N} \frac{\partial^2}{\partial x_i^2} - g \sum_{i,j=1}^{N} \delta(x_i - x_j). \tag{3.14}$$

The centre-of-mass (c.m.) solution for the bound state is (McGuire 1964):

$$\psi(x_1, x_2, \ldots x_N) = M \exp\left(-\tfrac{1}{4} g \sum_{i<j=1} |x_i - x_j|\right), \tag{3.15}$$

where M is the normalization constant.

The binding energy corresponding to this state is

$$E_N = -\tfrac{1}{48} g^2 N(N^2 - 1). \tag{3.16}$$

Now using the wavefunction (3.15) two main characteristics of the system, namely the form factor (density distribution) and momentum distribution can be found explicitly.

The density $\rho(x)$ is determined (Calogero and Degasperis 1975) in a closed form:

$$\rho(x) = \sum_{n=1}^{N-1} a_n \exp(-gnN|x|/2). \tag{3.17}$$

The Fourier transform of this expression gives the form factor:

$$F(q) = gN \sum_{n=1}^{N-1} \frac{a_n n}{q^2 + (gnN/2)^2}. \quad (3.18)$$

In (3.17) and (3.18) the coefficients a_n are:

$$a_n = \frac{1}{2} \frac{g(-1)^{n+1} n (N!)^2}{(N+n-1)!(N-n-1)!}. \quad (3.19)$$

It was shown (Amado 1976) that for a large number of particles N and $q^2 \gg \lambda^2 \gg (q/N)^2$:

$$F(q) = \frac{\pi q}{\lambda} [\sinh(\pi q/\lambda)]^{-1}, \quad (3.20)$$

where $\lambda = \tfrac{1}{2}gN$.

If $q^2 \gg \lambda^2$ the form factor has a regime of exponential decrease in q: $F \sim q e^{-\alpha q}$, in spite of the fact that F is a function of q^2.

It is interesting to note that this high-momentum behaviour of the form factor is of the same nature as that obtained in the Hartree approximation investigated by Calogero and Degasperis (1975).

The expression for the momentum distribution $n(q)$ in the schematic model is also obtained (Amado and Woloshyn 1977b), but it is far too complicated for direct analysis. It was shown, however, that at $N \gg 1$ and $q^2 \leq \tfrac{1}{4}\lambda^2$ the exact form of $n(q)$ approximates well to the Hartree result for the schematic model:

$$n(q) = [\cosh(\pi q/\lambda)]^{-2}. \quad (3.21)$$

This formula was extensively used (3.10b) for analysing the inclusive backward-proton production in the process of high-energy proton-nucleus scattering (Amado and Woloshyn 1976a).

Amado (1976) obtained the important result that the momentum distribution has the following power-law at high values of q,

$$n(q) \sim [\tilde{v}(q)]^2 \frac{1}{q^4}, \quad (3.22)$$

where $\tilde{v}(q)$ is the Fourier transform of the two-body potential. In the case of δ forces:

$$n(q) \sim 1/q^4. \quad (3.23)$$

It is not clear, however, whether it is q or q/N that must be large for this asymptotic regime.

The situation is quite different in the case of the form factor whose asymptotic value

$$F(q) \sim (\tilde{v}(q)/q^2)^{N-1} \qquad (3.24)$$

holds when q/N is large.

Such asymptotic behaviour of $F(q)$ is obtained by Narodetsky and Simonov (1975). It is also shown that an extra factor which increases the degree of q in the denominator of eqn (3.24) appears as a result of including the effects of the Pauli principle.

The validity of various forms of the momentum distribution in the asymptotic region remains a very intriguing question and needs further detailed study.

4

INDEPENDENT-PARTICLE MODEL DESCRIPTION

In this chapter the independent-particle model description of the nucleon momentum and density distributions is reviewed. The predictions for these two quantities in the case of non-interacting fermions and for $n(k)$ in the dilute Fermi-gas model are considered in Section 4.1. Analogous shell-model considerations are presented in Section 4.2. The Hartree–Fock approximation for infinite as well as for finite systems is discussed in Section 4.3. The main emphasis of this chapter is on the result presented in Section 4.4, namely that in any Hartree–Fock-type calculation it is not possible to reproduce simultaneously the density and momentum distributions in nuclei.

4.1. The Fermi-gas model

Some basic properties of fermion systems can be qualitatively described by means of the Fermi-gas model in which the particle wavefunctions are plane waves, i.e. the particles are moving freely in a volume of space Ω (the so-called independent-particle Fermi-gas). The total wavefunction of the system is simply constructed as a Slater determinant

$$\psi_\text{F}(\{r_i\}) = \hat{\mathcal{A}} \prod_{j=1}^{A} \varphi_{v_j}(r_j), \qquad (4.1)$$

where $\hat{\mathcal{A}}$ is the antisymmetrization operator and

$$\varphi_v(r) = \frac{1}{\Omega^{\frac{1}{2}}} e^{i k_v \cdot r} \qquad (4.2)$$

is the single-particle wavefunction for the state v and k_v is the nucleon momentum. Ω is the normalization volume. If it is taken as a cube with sides of length L then the relation between k and L is

$$k = \frac{2\pi}{L} (n_x, n_y, n_z), \qquad (4.3)$$

where n_x, n_y, and n_z are positive or negative integers.

INDEPENDENT-PARTICLE MODEL DESCRIPTION

The kinetic energy of the single-particle state is $\hbar^2 k_\nu^2/2m$. Using the wavefunction eqns (4.1) and (4.2) and the Hamiltonian

$$H = -\frac{\hbar^2}{2m}\sum_i \nabla_i^2, \qquad (4.4)$$

one can calculate all the characteristics of the system.

Thus the energy of the system is given by (see also eqn (1.37)):

$$E_0 = \langle \psi_F | -\frac{\hbar^2}{2m}\sum_i \nabla_i^2 | \psi_F \rangle$$

$$= -\frac{\hbar^2}{2m}\int [\nabla^2 \rho(r, r')]_{r'=r}\, dr, \qquad (4.5)$$

where the one-body density matrix $\rho(r, r')$ for this case is (1.25):

$$\rho(r, r') = \frac{4}{\Omega}\sum_{k<k_F} e^{-i\mathbf{k}\cdot\mathbf{r}'}e^{i\mathbf{k}\cdot\mathbf{r}}. \qquad (4.6)$$

The momentum of the last occupied state is labelled by k_F (the Fermi-momentum). Using the rule:

$$\sum_k \to \frac{\Omega}{(2\pi)^3}\int d\mathbf{k}, \qquad (4.7)$$

we can obtain

$$\rho(r, r') = 3\rho_0 \frac{j_1(k_F |r' - r|)}{k_F |r' - r|}, \qquad (4.8)$$

where $\rho_0 = A/\Omega = $ constant. Then from eqn (4.5) we obtain the energy of the system:

$$E_0 = \tfrac{3}{5}A \frac{\hbar^2 k_F^2}{2m}. \qquad (4.9)$$

The average energy per nucleon is given by

$$\frac{E_0}{A} = \frac{3}{5}\frac{\hbar^2 k_F^2}{2m}. \qquad (4.9a)$$

Obviously the density of the free Fermi-gas system is constant which can be formally seen from (4.8)

$$\rho(r) = \rho(r, r')|_{r'=r} = \rho_0. \qquad (4.10)$$

Using eqns (4.8) or (4.6) the nucleon momentum distribution of the system

$$n(\mathbf{k}) = \int e^{i\mathbf{k}\cdot(\mathbf{r}-\mathbf{r}')}\rho(r, r')\, dr\, dr', \qquad (4.11)$$

becomes:

$$n(\mathbf{k}) = 4\Omega\theta(k_F - |\mathbf{k}|), \quad (4.12)$$

where θ is the step function (4.15). Taking into account the normalization condition:

$$\int n(\mathbf{k}) \frac{d^3k}{(2\pi)^3} = A, \quad (4.13)$$

the important relation between the Fermi-momentum and the number of particles follows:

$$k_F = \left(\frac{3\pi^2}{2}\frac{A}{\Omega}\right)^{\frac{1}{3}} = \left(\frac{3\pi^2}{2}\rho_0\right)^{\frac{1}{3}}. \quad (4.14)$$

The factor 4 in (4.12) takes into account the spin-isospin degeneracy (in each state with momentum $\hbar\mathbf{k}$, 4 particles can be placed).

The step function

$$\theta(k_F - |\mathbf{k}|) = \begin{cases} 1, & |\mathbf{k}| < k_F \\ 0, & |\mathbf{k}| > k_F, \end{cases} \quad (4.15)$$

represents the occupation numbers (often identified with the momentum distribution $n(k)$). All the states with $|\mathbf{k}| < k_F$ within the Fermi-sphere with radius k_F are occupied (see Fig. 4.1) and all the higher states are unoccupied.

Since the kinetic energy of any occupied single-particle state is

$$E_{k_\nu} = \frac{\hbar^2 k_\nu^2}{2m}, \quad (4.16)$$

the energy for the last occupied state (the Fermi-energy) is given by the expression

$$E_{k_F} \equiv E_F = \frac{\hbar^2 k_F^2}{2m}. \quad (4.17)$$

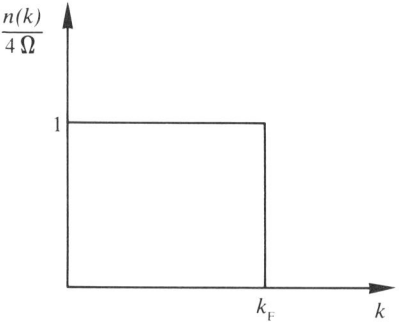

FIG. 4.1. Fermi-gas nucleon momentum distribution.

The next natural step is the development of the Fermi-gas model with inclusion of rather simple (schematic) interactions between fermions.

Here we shall consider qualitatively following Czyż and Gottfried (1961) the change in the momentum distribution caused by a certain type of repulsive interaction between the particles. It is clear that in principle the inclusion of the interaction could lead to the partial depletion of the Fermi-sea and to the partial occupation of the states outside it.

Using the techniques developed by Galitski (1958) and Galitski and Migdal (1958) for a dilute Fermi-gas with repulsive interaction to second-order parameter $k_F a$ ($a > 0$ being the scattering length) the following distribution is obtained:

$$n(k > k_F) = \delta n_+(k), \tag{4.18}$$

$$n(k < k_F) = 1 + \delta n_-(k), \tag{4.19}$$

where

$$\delta n_{\mp}(k) = \mp \frac{a^2}{4\pi^4 m^2} \int d\boldsymbol{p}\, d\boldsymbol{\kappa}\, \frac{\theta_{\pm}(p)\Lambda_{\mp}(\kappa)}{(P^2/4m + \kappa^2/m - E_p^0 - E_k^0)}. \tag{4.20}$$

In (4.20)

$$\boldsymbol{P} = \boldsymbol{k} + \boldsymbol{p}, \tag{4.21}$$

$$\Lambda_{\mp}(\boldsymbol{\kappa}) = \theta_{\mp}(\tfrac{1}{2}\boldsymbol{P} + \boldsymbol{\kappa})\theta_{\mp}(\tfrac{1}{2}\boldsymbol{P} - \boldsymbol{\kappa}), \tag{4.22}$$

$$\theta_-(k) = \begin{cases} 1, & k < k_F \\ 0, & k > k_F \end{cases}; \quad \theta_+(k) = 1 - \theta_-(k), \tag{4.23}$$

$$E_p^0 = p^2/2m. \tag{4.24}$$

In the vicinity of k_F there is a discontinuity of $n(k)$ of an amount $[1 - 4(k_F a/\pi)^2 \ln 2]$ at the Fermi-surface. It is interesting to note that $n(k)$ varies as k^{-4} when $k/k_F \to \infty$. This k^{-4} long-tail momentum distribution behaviour is valid when the inequalities $k \gg k_F$ and $ka < 1$ are satisfied simultaneously. Note that this asymptotic result is in agreement with the result of the schematic model presented in Section 3.3. One can hope that the results obtained for $n(k)$ give a good indication of the role of the nucleon–nucleon hard core if the parameter a is of the order of core size, $a \cong 0.4$ fm.

The problem of discontinuity of the momentum distribution is examined for Fermi-liquids by Migdal (1957, 1967). It is shown that the discontinuity in the momentum distribution at $k = k_F$ is an inherent consequence of an arbitrary interaction between particles in an infinite system.

The problem of the nucleon momentum distribution of a dilute Fermi-gas has also been considered by Belyakov (1961) and by Sartor and Mahaux (1980a). In (Belyakov 1961) the momentum distribution is obtained using perturbation theory up to the term of order $(an^{\frac{1}{3}})^2$, where n is the number of particles per unit volume and a is the s-wave scattering amplitude. The approach and results of Sartor and Mahaux (1980a) are discussed in Section 5.2.

4.2. The shell-model

The main assumption of the shell-model is that the nucleons move independently of each other in a common average field. The single-particle potential related to this field acting on a separate nucleon can be considered as created by all the other nucleons. Formally it can be introduced as follows. If the Hamiltonian of the system of A nucleons is of the form:

$$\hat{H} = -\frac{\hbar^2}{2m}\sum_i \nabla_i^2 + \sum_{i<j} v_{ij} \equiv T + V, \quad (4.25)$$

where v_{ij} is the nucleon–nucleon interaction, it is possible to introduce the average (mean) potential U by writing (4.25) in the form:

$$\hat{H} = \hat{T} + U + \hat{V} - U \equiv \hat{H}_0 + \hat{W}, \quad (4.26)$$

where

$$\hat{H}_0 = \hat{T} + U, \quad \hat{W} = \hat{V} - U. \quad (4.27)$$

The Hamiltonian of the shell-model is taken to be $\hat{H}_0$,

$$\hat{H} \simeq \hat{H}_0. \quad (4.28)$$

The mean field U is now chosen so that the residual interaction $\hat{W}$ can be considered as a small perturbation. This greatly simplifies the many-body problem and there remains only the problem of specifying U.

Usually several simple forms of this potential are chosen:

i) the isotropic harmonic-oscillator potential:

$$U(r) = -U_0 + \tfrac{1}{2}m\omega^2 r^2; \quad (4.29)$$

ii) the spherical potential well

$$U(r) = \begin{cases} -U_0, & r < R \\ 0, & r > R; \end{cases} \quad (4.30)$$

iii) the Saxon–Woods potential

$$U(r) = -U_0 \frac{1}{1 + e^{(r-R)/a}}. \quad (4.31)$$

In each case the parameters of these potentials are fitted by means of a

INDEPENDENT-PARTICLE MODEL DESCRIPTION

comparison with the experimental data on single-particle energies, density distribution, and radii.

The shell-model wavefunctions $\psi_{SM}(\{r_i\})$ which is the solution of the many-particle Schrödinger equation

$$\hat{H}_0 \psi = E\psi \tag{4.32}$$

has the form of a Slater determinant,

$$\psi_{SM}(r_1, \ldots r_A) = \frac{1}{\sqrt{A!}} \det|\varphi_i(r_j)|, \quad i,j = 1, \ldots A. \tag{4.33}$$

The single-particle wavefunctions $\varphi_i(r)$ are solutions of the one-particle equation

$$\left[-\frac{\hbar^2}{2m}\nabla^2 + U(r)\right]\varphi(r) = \varepsilon\varphi(r), \tag{4.34}$$

where $U(r)$ is the one-particle potential (for instance, one of those defined by eqns (4.29), (4.30), and (4.31)). ε the eigenvalue of the equation is the one-particle energy.

Since a detailed description of the single-particle properties is not our principal concern we do not include the important spin-orbit interaction.

The ground-state energy $E = E_0$ is a sum of all single-particle energies

$$E_0 = \sum_{i=1}^{A} \varepsilon_i. \tag{4.35}$$

Since we have the ground-state wavefunction $\psi_{SM}(\{r_i\})$ all the bulk characteristics can be obtained, namely the local density $\rho(r)$ and momentum distribution $n(k)$, the radii, and the various moments of nuclei.

For our purposes we need the distributions related to the one-body density matrix corresponding to ψ_{SM}, see eqn (4.33):

$$\rho(r, r') = \sum_{i=1}^{A} \varphi_i^*(r')\varphi_i(r), \tag{4.36}$$

namely the local density distribution

$$\rho(r) = \sum_{i=1}^{A} |\varphi_i(r)|^2, \tag{4.37}$$

and the nucleon momentum distribution

$$n(k) = \rho(k, k' = k) = \sum_{i=1}^{A} |\tilde{\varphi}_i(k)|^2, \tag{4.38}$$

where $\tilde{\varphi}(k)$ is the Fourier transform of $\varphi(r)$.

In the case of the harmonic-oscillator shell-model (4.29) we give the explicit expressions for the density and momentum distributions in ^{4}He and ^{16}O:

^{4}He:
$$\rho(r) = \frac{4}{\pi^{\frac{3}{2}}} \alpha^3 e^{-\alpha^2 r^2}, \tag{4.39}$$

$$n(k) = \frac{32\pi^{\frac{3}{2}}}{\alpha^3} e^{-k^2/\alpha^2}, \tag{4.40}$$

^{16}O:
$$\rho(r) = \frac{4\alpha^3}{\pi^{\frac{3}{2}}} (1 + 2\alpha^2 r^2) e^{-\alpha^2 r^2}, \tag{4.41}$$

$$n(k) = \frac{32\pi^{\frac{3}{2}}}{\alpha^3} \left(1 + \frac{2k^2}{\alpha^2}\right) e^{-k^2/\alpha^2}, \tag{4.42}$$

where

$$\alpha = \left(\frac{m\omega}{\hbar}\right)^{\frac{1}{2}}, \qquad \hbar\omega \approx 41 A^{-\frac{1}{3}} \text{ MeV}, \tag{4.43}$$

$$\int \rho(r) \, d\mathbf{r} = A, \qquad \int n(\mathbf{k}) \frac{d\mathbf{k}}{(2\pi)^3} = A. \tag{4.44}$$

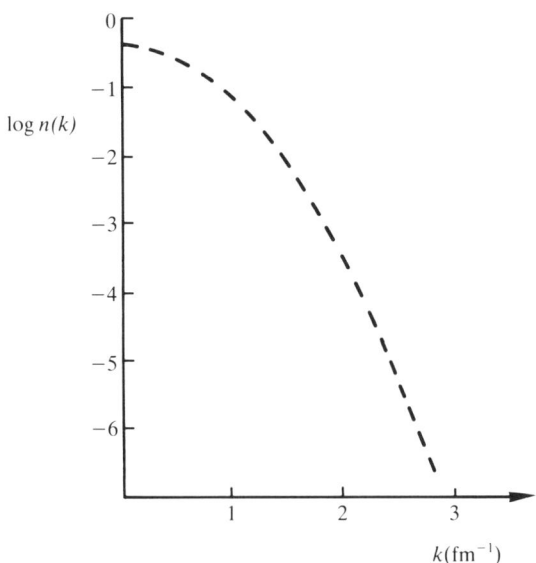

FIG. 4.2. Momentum distribution of ^{4}He from shell-model calculations with harmonic-oscillator potential (Dal Rì et al. 1982).

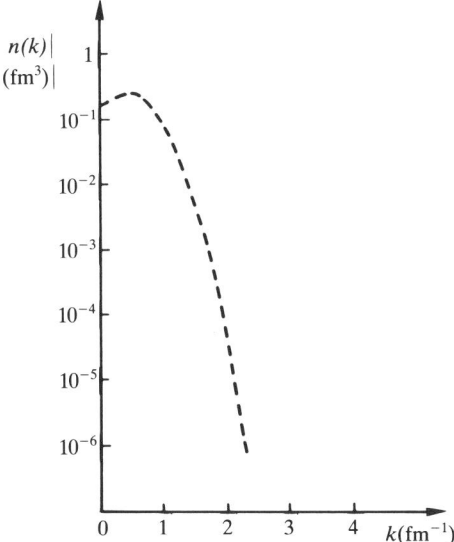

FIG. 4.3. Momentum distribution of ^{16}O from shell-model calculations with harmonic-oscillator potential (Małecki and Picchi 1973).

The characteristic feature of the local density distribution in the case of the harmonic-oscillator potential is the Gaussian decrease at $r \to \infty$. This distribution drastically deviates from the realistic distributions obtained in different experiments and is in contradiction with the exponential decrease of $\rho(r)$ characteristic of real nuclei.

The nucleon momentum distribution for ^{4}He and ^{16}O eqns (4.40) and (4.42) are given in Figs. 4.2 and 4.3. The main feature of these distributions $n(k)$ is the steep slope when k increases.

4.3. The Hartree–Fock approximation

Infinite systems

In order to facilitate further generalization (Chapter 5) it is convenient to introduce the Hartree–Fock approximation using the Green-function method (see also Section 1.4).

For infinite systems due to space- and time-translational invariance the one-particle Green function in momentum and energy representations has the form (1.95):

$$G(k, E) = \frac{1}{E - k^2/2m - \Sigma(k, E)}, \qquad (4.45)$$

and obeys the Dyson equation

$$G(k, E) = G^0(k, E) + G^0(k, E)[\Sigma(k, E) - U(k)]G(k, E), \quad (4.46)$$

where $G^0(k, E)$ is the one-particle Green function corresponding to a non-interacting system, for which

$$\Sigma^{(0)}(k, E) = U(k). \quad (4.47)$$

When the nucleon–nucleon interaction ϑ permits perturbation expansion in powers of its strength, the mass operator can be written as (Mahaux et al. 1985; Jeukenne et al. 1976)

$$\Sigma(k, E) = \Sigma_{1a}(k) + \Sigma_{1b}(k, E) + \ldots . \quad (4.48)$$

The first-order term $\Sigma_{1a}(k)$ is called the Hartree–Fock potential and its explicit form is:

$$\Sigma_{1a}(k) = V_{HF}(k) = \sum_j \theta(k_F - |j|)\{\langle \mathbf{k}, \mathbf{j}| \vartheta |\mathbf{k}, \mathbf{j}\rangle$$
$$- \langle \mathbf{k}, \mathbf{j}| \vartheta |\mathbf{j}, \mathbf{k}\rangle\}, \quad (4.49)$$

where

$$|\mathbf{k}, \mathbf{j}\rangle = \exp\{i(\mathbf{k} \cdot \mathbf{r}_1 + \mathbf{j} \cdot \mathbf{r}_2)\}. \quad (4.50)$$

It is seen that eqn (4.49) contains the direct (Hartree) term:

$$V_H(k) = \sum_j \theta(k_F - |j|)\langle \mathbf{k}, \mathbf{j}| \vartheta |\mathbf{k}, \mathbf{j}\rangle, \quad (4.51)$$

and the exchange (Fock) term:

$$V_F(k) = -\sum_j \theta(k_F - |j|) \langle \mathbf{k}, \mathbf{j}| \vartheta |\mathbf{j}, \mathbf{k}\rangle \quad (4.52)$$

related to the Pauli principle.

As can be seen the Hartree–Fock potential is static, i.e. is independent of the energy and so local in time. It is also momentum dependent which means it is non-local in space.

The corresponding Green function in the Hartree–Fock approximation is now:

$$G_{1a}(k, E) = \frac{\theta(k_F - k)}{E - \hat{e}(k) - i\eta} + \frac{\theta(k - k_F)}{E - \hat{e}(k) + i\eta}, \quad (4.53)$$

where

$$\hat{e}(k) = k^2/2m + \Sigma_{1a}(k). \quad (4.54)$$

The momentum distribution is simply related to the Green function $G(k, E)$:

$$n(k) = -\frac{i}{2\pi} \sum_k \int_C dE\, G(k, E), \quad (4.55)$$

INDEPENDENT-PARTICLE MODEL DESCRIPTION

(see also eqn (1.101)). The integration contour C is formed of the real axis and is closed in the upper half plane.

Using (4.53) one can obtain the momentum distribution in the Hartree–Fock approximation:

$$n_{1a}(k) \equiv n_{HF}(k) = \theta(k_F - k), \qquad (4.56)$$

which coincides with the momentum distribution for the free Fermi-gas; this means that in this approximation the ground-state motion is not correlated.

For the binding energy the well-known Hartree–Fock expression can be obtained.

$$\mathcal{B}_{HF} = \langle T \rangle + \tfrac{1}{2} \sum_{k,j} \theta(k_F - |k|) \theta(k_F - |j|) \langle k, j | \vartheta | k, j - j, k \rangle, \qquad (4.57)$$

where $\langle T \rangle = \sum_k \theta(k_F - k) k^2/2m$ is the kinetic energy of the non-interacting ground state.

Finite nucleon systems

For a finite system the analogue of eqn (4.45) is:

$$[E + \nabla_r^2/2m] G(\mathbf{r}, \mathbf{r}', E) = \delta(\mathbf{r} - \mathbf{r}') + \int d\mathbf{r}' \Sigma(\mathbf{r}, \mathbf{r}'; E) G(\mathbf{r}, \mathbf{r}', E), \qquad (4.58)$$

where the one-body Green-function Lehmann representation is:

$$G(\mathbf{r}, \mathbf{r}', E) = \sum_h \frac{\varphi_h(\mathbf{r}) \varphi_h^*(\mathbf{r}')}{E - E_h - i\eta} + \sum_p \frac{\varphi_p(\mathbf{r}) \varphi_p^*(\mathbf{r}')}{E - E_p + i\eta}, \qquad (4.59)$$

with

$$\varphi_h(\mathbf{r}) = \langle \psi_h^{(A-1)} | a(\mathbf{r}) | \psi_0^{(A)} \rangle, \qquad (4.60)$$

$$\varphi_p(\mathbf{r}) = \langle \psi_0^{(A)} | a(\mathbf{r}) | \psi_p^{(A+1)} \rangle. \qquad (4.61)$$

In eqns (4.60) and (4.61) $a(\mathbf{r})$ is the particle annihilation operator, $\psi_h^{(A-1)}$ is the eigenstate of the Hamiltonian of the $(A-1)$ particle system with energy $E_h^{(A-1)}$, and $\psi_p^{(A+1)}$ is an eigenstate of the $(A+1)$ particle system with energy $E_p^{(A+1)}$.

In eqn (4.59):

$$E_p = E_p^{(A+1)} - E_0^{(A)}, \qquad (4.62)$$

$$E_h = E_0^{(A)} - E_h^{(A-1)}, \qquad (4.63)$$

and $E_0^{(A)}$ is the energy corresponding to $\psi_0^{(A)}$.

From eqns (4.59) and (4.58) the single-particle wavefunction can be obtained (Mahaux et al. 1985):

$$[E_p + \nabla^2/2m]\varphi_p(r) - \int dr' \Sigma(r, r'; E_p)\varphi_p(r') = 0. \quad (4.64)$$

In the first approximation ($\Sigma \to V_{HF}$) the self-consistent system of Hartree–Fock equations is obtained

$$-\frac{\nabla^2}{2m}\varphi_k + V_{HF}\varphi_k = \varepsilon_k\varphi_k, \quad (4.65)$$

where

$$V_{HF}\varphi_k = V_H(r)\varphi_k - \int V_F(r, r')\varphi_k(r') dr'. \quad (4.66)$$

Here the Hartree and Fock contributions are:

$$V_H(r) = \sum_j n_j \langle \varphi_j(r') | \vartheta(r, r') | \varphi_j(r') \rangle, \quad (4.67)$$

$$V_F(r, r') = \sum_j n_j \varphi_j^*(r') \vartheta(r, r') \varphi_j(r), \quad (4.68)$$

and $n_j = 1$ for the A deepest bound single-particle orbits and 0 for other orbits.

The Hartree–Fock basis $\{\varphi_j\}$ minimizes the ground-state expectation value of the Hamiltonian

$$E_{HF} = \langle \psi_{HF} | H | \psi_{HF} \rangle, \quad (4.69)$$

where ψ_{HF} is a Slater determinant built from the A deepest bound orbitals φ_j.

In principle, the mean-field approaches like the Hartree–Fock one are fundamental in the sense that they enable many nuclear properties to be calculated from the nucleon–nucleon interaction. In practice, however, this is not possible because of the presence of a hard core and also because the large basis required for realistic calculations leads to great numerical complexity. These difficulties are usually overcome by introducing some phenomenological elements into the calculations. The set of basis states must be truncated and the parameters optimized for the basis chosen. Another phenomenological element is that an effective interaction is used with some parameters that are adjusted to fit selected experimental data. In terms of beyond-Hartree–Fock approaches (see Chapter 5) the interaction is renormalized by a G-matrix expansion in the Brueckner–Hartree–Fock method, or alternatively some of the higher order many-body terms are taken into account by introducing a density dependence. The resulting velocity-dependent and density-dependent phenomenological interactions have been widely used to analyse nuclear-structure data (extended reviews of Hartree–Fock calculations are presented in (Svenne 1979; Bartz et al. 1982)).

Practical calculations in Hartree–Fock approaches are carried out using effective interactions ϑ whose parameters are adjusted by comparison of the energy E_{HF}, the density distribution

$$\rho_{HF}(r) = \sum_j n_j |\varphi_j(r)|^2, \tag{4.70}$$

and the Fermi-energy

$$\varepsilon_F = \tfrac{1}{2}[E_0^{(A+1)} - E_0^{(A-1)}], \tag{4.71}$$

with the experimental data.

The nucleon momentum distribution in the Hartree–Fock approximation is of the form

$$n_{HF}(k) = \sum_j n_j |\tilde{\varphi}_j(k)|^2, \tag{4.72}$$

where $\tilde{\varphi}_j(k)$ is the Fourier transform of $\varphi_j(r)$.

The nucleon density and momentum distributions can be obtained by means of the one-body density matrix which has the following form in the Hartree–Fock approximation:

$$\rho_{HF}(r, r') = \sum_j n_j \varphi_j^*(r') \varphi_j(r). \tag{4.73}$$

The Hartree–Fock calculations may be simplified in practice using the Skyrme-type effective interaction. This is expressed in terms of six parameters that are adjusted to fit the volume, surface, symmetry, Coulomb, and spin-orbit energies of nuclei. The energy in this case is obtained in the simple form:

$$E_{HF} = \int dr\, \mathcal{H}(r), \tag{4.74}$$

where the energy density $\mathcal{H}(r)$ depends upon $\rho_{HF}(r)$ and on the kinetic-energy density $\tau_{HF}(r)$:

$$\tau_{HF}(r) = \frac{1}{2m} \sum_j n_j |\nabla \varphi_j|^2. \tag{4.75}$$

Numerous calculations have been carried out in the Hartree–Fock approximation with different forces and especially with Skyrme-type effective forces. In general they show good results for the bulk nuclear properties, such as binding energies and mean-square radii. As for the nucleon density distribution the results are in good agreement with the

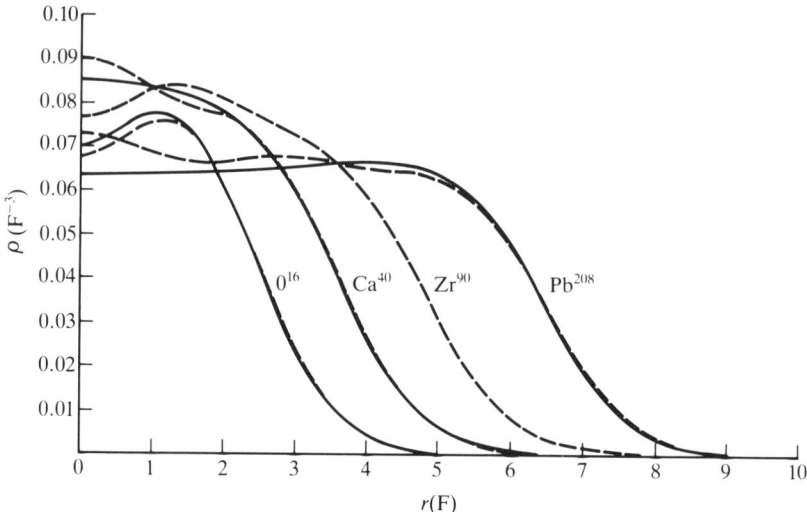

FIG. 4.4. Empirical (solid lines) and theoretical (dashed lines) charge distributions for ^{16}O, ^{40}Ca, ^{90}Zr, and ^{208}Pb. The theoretical curves represent the density-dependent Hartree-Fock calculations of Negele (1970).

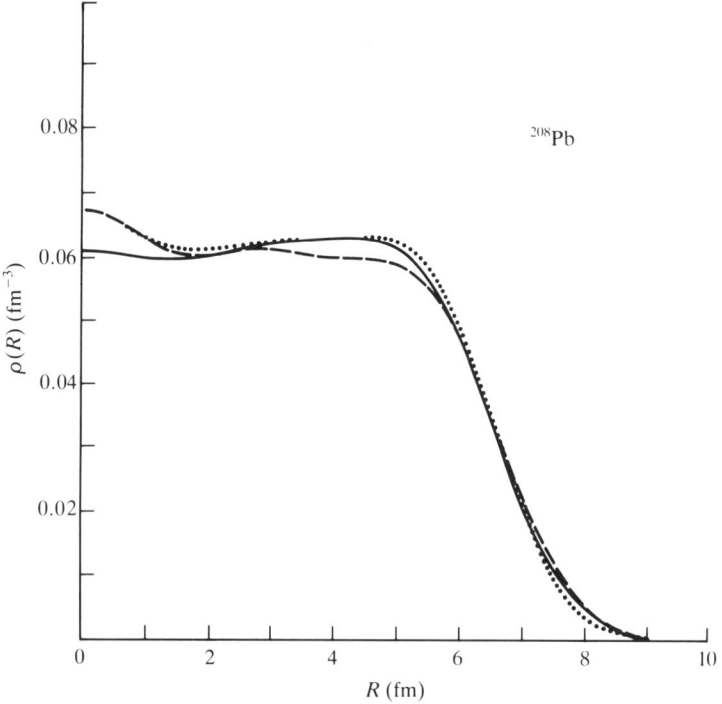

FIG. 4.5. The Hartree-Fock calculations of Vautherin and Brink (1972) of the ^{208}Pb charge distribution using Skyrme-I (short dash) and Skyrme-II (long dash) set of parameters in comparison with experimental data (solid line).

experimental data on electron and proton scattering on nuclei, although some oscillations in the inner region of the density distribution are rather more prominent than are found experimentally. This can be seen in the following figures. Fig. 4.4 shows a comparison between the experimental data for charge distributions of ^{16}O, ^{40}Ca, ^{90}Zr, and ^{208}Pb and the results of the density-dependent Hartree–Fock calculations of Negele (1970). Mentioned also should be the Hartree–Fock calculations of Campi and Sprung (1972) with GO interactions and those of Dechargè et al. (1978) with D1 interactions. The experimental charge distribution of ^{208}Pb is compared in Fig. 4.5 with Hartree–Fock calculations of Vautherin and Brink (1972) using Skyrme-I and Skyrme-II sets of parameters and in Fig. 4.6 with density-dependent Hartree–Fock calculations of Gogny (1979).

The situation with the nucleon momentum distribution, however, is much worse. The results for $n(k)$ are quite similar to those obtained in the shell-model calculations in that they exhibit a steep fall-off for large momenta. In Figs. 4.7 and 4.8 we show the results for $n(k)$ in the cases of

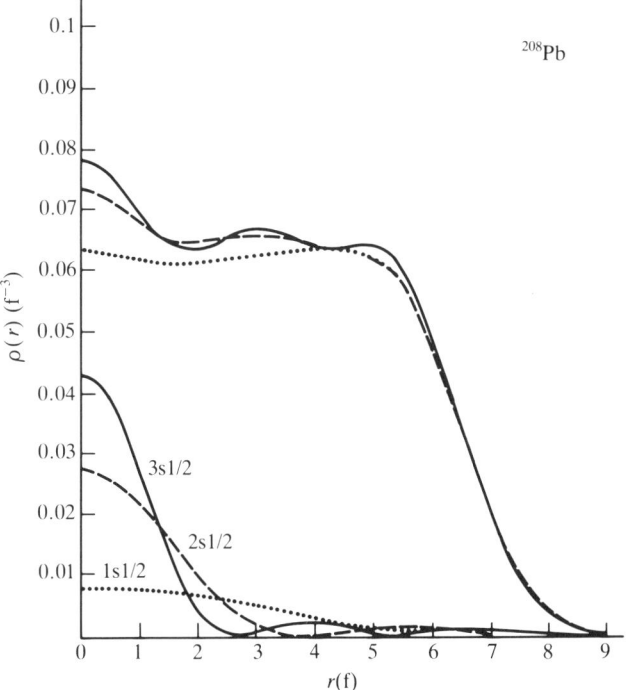

FIG. 4.6. The proton (solid line) and charge (dashed line) distributions of ^{208}Pb calculated in the density-dependent Hartree-Fock method of Gogny (1979) in comparison with the experimental data (...). The contributions of $1s_{\frac{1}{2}}$, $2s_{\frac{1}{2}}$, and $3s_{\frac{1}{2}}$ are drawn separately.

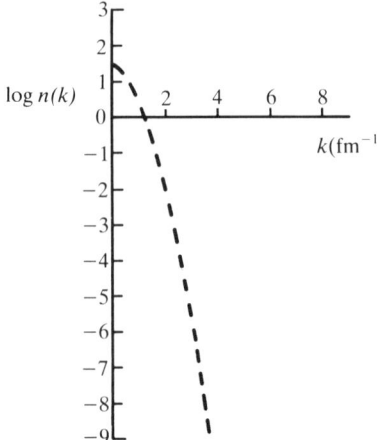

FIG. 4.7. Momentum distribution of ^{4}He from the Hartree-Fock calculations with Reid soft-core potential (Zabolitzky and Ey 1978).

^{4}He and ^{40}Ca nuclei calculated using Reid soft-core forces (Zabolitzky and Ey 1978). The main reason for this behaviour of $n(k)$ is related to the fact that the determinant Hartree–Fock ground-state wavefunction ψ_{HF} including the Pauli correlations does not take into account the important effects of the short-range dynamical correlations. As was quantitatively shown in Chapter 3, the short-range repulsive features of the nucleon–nucleon forces could be responsible for the high-momentum behaviour of the momentum distribution.

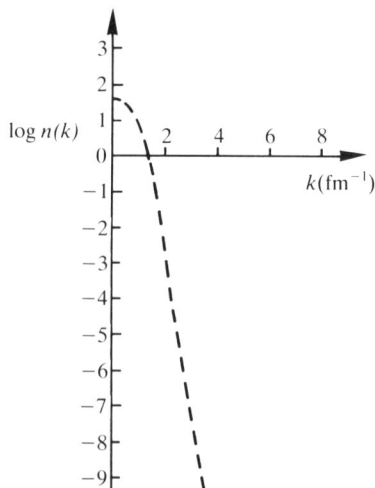

FIG. 4.8. Momentum distribution of ^{16}O from the Hartree-Fock calculations with Reid soft-core potential (Zabolitzky and Ey 1978).

4.4. A remark on the reliability of Hartree–Fock predictions

As discussed in Section 4.3 the form and parameters of the nucleon-nucleon forces used in the Hartree–Fock calculations are established by a comparison of the calculated binding energies, Fermi-energy, and radii with their experimental values, as well as with the nuclear-matter characteristics (energy-saturation curve). In general, reasonable results for the diagonal elements of the one-body density matrix (i.e. the local density ρ) are obtained for a wide range of nuclei. We note that in the Hartree–Fock calculations the bulk characteristics sensitive to the density distribution are most dominant, but they are less sensitive to the characteristics largely determined by the non-diagonal elements of the one-body density matrix. These characteristics (for instance, the nucleon momentum distribution) are obtained only as a final result of the Hartree–Fock calculations. The question arises however about the predictive power of the Hartree–Fock approaches for the characteristics related to the non-diagonal elements of $\rho(r, r')$.

As shown by Kutzelnigg and Smith (1964), Kobe (1969), and Jaminon et al. (1985a) (see also Section 1.5) a reasonable criterion for the proximity of the one-body density matrix $\rho_0(r, r')$ in the Hartree–Fock approximation to the exact solution for the one-body density matrix $\rho(r, r')$ is that the mean-square deviation of these two matrices should be minimal:

$$\mathrm{Tr}[(\rho - \rho_0)^2] = \text{minimum}. \qquad (4.76)$$

This condition is satisfied if $\rho_0(r, r')$ is constructed from the Slater determinant expressed in terms of natural orbitals.

Solving the many-body problem in the Hartree–Fock approximation gives the one-body density matrix $\rho_0^{\mathrm{HF}}(r, r')$ which generally differs from the $\rho_0(r, r')$ which satisfies the relation (4.76). As shown by Jaminon et al. (1985a) using the Hartree–Fock fitting procedure to reproduce reasonably well the diagonal elements of $\rho(\rho(r, r) \simeq \rho_0^{\mathrm{HF}}(r, r))$, which give the density distribution, inevitably increases the deviation between the non-diagonal elements of these two matrices ($\rho(r, r')$ and $\rho_0^{\mathrm{HF}}(r, r')$ at $r \neq r'$). Since the non-diagonal elements of the one-body density matrix are related to the nucleon momentum distribution, it is impossible to reproduce the latter by conventional Hartree–Fock procedure. In other words, one cannot hope to reproduce simultaneously in the Hartree–Fock approximation both the nucleon density distribution (diagonal elements of $\rho(r, r')$) and nucleon momentum distribution ($\sim$non-diagonal elements of $\rho(r, r')$). This conclusion must be kept in mind when analysing the experiments which probe the momentum distribution in nuclei.

5

BEYOND HARTREE–FOCK METHODS

In this chapter we outline the perturbation expansion of the mass operator (Section 5.1), the hole-line expansion of the mass operator, the Brueckner-Hartree–Fock approach (Section 5.2), and the exp(S) method (Section 5.3). Particular attention is paid to the effects of the beyond-Hartree–Fock terms on the distributions of interest.

5.1. The perturbation expansion of the mass operator

The necessity of developing methods beyond the scope of the Hartree-Fock approximation which was partially discussed in Section 4.4 requires consideration of the higher terms in the mass-operator expansion (Jeukenne *et al.* 1976)

$$\Sigma(k, E) = \Sigma_{1a}(k) + \Sigma_{1b}(k, E) + \Sigma_{1c}(k, E) \\ + \Sigma_2(k, E) + \Sigma_3(k, E) + \ldots \quad (5.1)$$

It has been pointed out (4.49) that the first term $\Sigma_{1a}(k)$, which is the first-order term of the nucleon-nucleon interaction, corresponds to the Hartree–Fock approximation. The following terms are of second and higher order in ϑ.

As we have seen (4.56), the momentum distribution in the Hartree–Fock approximation is:

$$n_{1a}(k) \equiv n_{HF}(k) = \theta(k_F - k). \quad (5.2)$$

In order to find the corrections to the Hartree–Fock momentum distribution n_{HF} we have to consider the terms $\Sigma_{1b}(k, E)$ and $\Sigma_2(k, E)$ which are of second order in ϑ and use the definition of the momentum distribution (4.55).

The expressions for $\Sigma_{1b}(k, E)$ and $\Sigma_2(k, E)$ are given by:

$$\Sigma_{1b}(k, E) = \tfrac{1}{2} \sum_{j,a,b} \theta(k_F - |j|)\theta(|a| - k_F)\theta(|b| - k_F) \\ \times \frac{|\langle k, j| \vartheta | a, b-b, a \rangle|^2}{E - \hat{e}(j) - \hat{e}(a) - \hat{e}(b) + i\eta}, \quad (5.3)$$

$$\Sigma_2(k, E) = \frac{1}{2} \sum_{l,j,a} \theta(k_F - |j|)\theta(k_F - |l|)\theta(|a| - k_F)$$

$$\times \frac{|\langle j, l| \vartheta |k, a-a, k\rangle|^2}{E - \hat{e}(a) - \hat{e}(j) - \hat{e}(l) - i\eta}, \quad (5.4)$$

where

$$\hat{e}(d) = d^2/2m + \hat{U}(d), \quad (5.5)$$

with the auxiliary potential $\hat{U}(d)$ chosen so as to improve the convergence of the expansion.

The expressions (5.3) and (5.4) as well as the term Σ_{1a} are graphically represented in Fig. 5.1 for the case of particle states (the upward- and downward-pointing arrows represent particles and holes respectively).

The graph (1b) is called a 'core polarization' graph and (2) a ground-state 'correlation graph'.

The corresponding leading corrections to the nuclear-matter Hartree–Fock momentum distribution are of the form:

$$n_{1b}(k) = -\tfrac{1}{2} \sum_{l,c,d} \theta(k_F - |l|)\theta(|c| - k_F)\theta(|d| - k_F)$$

$$\times \frac{|\langle k, l| \vartheta |c, d-d, c\rangle|^2}{[\hat{e}(k) + \hat{e}(l) - \hat{e}(c) - \hat{e}(d)]^2}, \quad \text{for } k < k_F, \quad (5.6)$$

$$n_2(k) = \tfrac{1}{2} \sum_{a,j,l} \theta(|a| - k_F)\theta(k_F - |j|)\theta(k_F - |l|)$$

$$\times \frac{|\langle k, a| \vartheta |j, l-l, j\rangle|^2}{[\hat{e}(k) + \hat{e}(a) - \hat{e}(j) - \hat{e}(l)]^2}, \quad \text{for } k > k_F. \quad (5.7)$$

It follows that up to second order, for $k < k_F$

$$n_1(k) = n_{HF}(k) + n_{1b}(k). \quad (5.8)$$

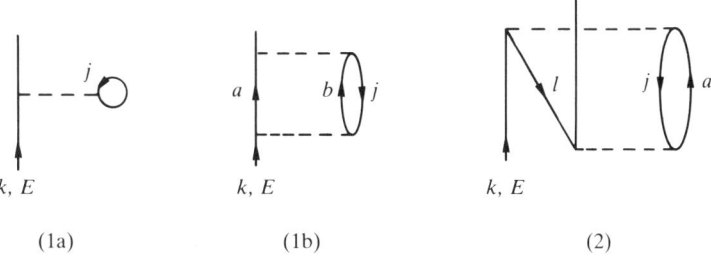

FIG. 5.1. The graphs corresponding to Σ_{1a}, Σ_{1b}, and Σ_2 from the perturbation expansion in the case of particle states (Jeukenne et al. 1976).

As we can see, the consideration of the higher-order terms beyond the Hartree–Fock approximations leads to the depletion of the Fermi-sea (the term n_{1b}) and to the appearance of $n(k)$ components above k_F (the term $n_2(k)$).

In the paper of Orland and Schaeffer (1978) a simple phenomenological model is suggested in order to find an explicit approximate form of the mass operator, using the dispersion relation for the latter (1.97). The imaginary part of the mass operator is suitably parametrized by means of the two-body nucleon–nucleon scattering data. Using the dispersion relation, Orland and Schaeffer (1978) obtain the real part of the first two terms of the mass operator, after the Hartree–Fock term, Σ_1 and Σ_2 (corresponding to 'core polarization' and 'correlation graph' respectively):

$$\operatorname{Re}\Sigma_{1,2}(\omega) = \frac{\alpha\beta^2}{2\pi}\left(\frac{k_F^2}{2m}\right)^2\left(\frac{\varepsilon_F - \omega}{k_F^2/2m}\right)\left[1 \pm \frac{\varepsilon_F - \omega}{\beta k_F^2/2m}\exp\left[\pm\frac{\varepsilon_F - \omega}{\beta k_F^2/2m}\right]\right.$$
$$\left. \times \operatorname{Ei}\left(\mp\frac{\varepsilon_F - \omega}{\beta k_F^2/2m}\right)\right], \quad (5.9)$$

where Ei is the integral exponential function,

$$\alpha = \frac{m}{(2\pi)^2}\sigma\left[\frac{1}{2}\left(\frac{k^2}{2m} + \frac{k_F^2}{2m}\right)\right], \quad \beta \simeq 3, \quad (5.10)$$

and σ is the total nucleon–nucleon cross-section averaged over spin and isospin.

The graphs corresponding to Σ_1 (Σ_{1b} for the hole states in eqn (5.1)) and Σ_2 are illustrated in Fig. 5.2.

After determination of the mass operator the Green function is completely defined within the accepted approximation, and using the relation between the momentum distribution and the Green function

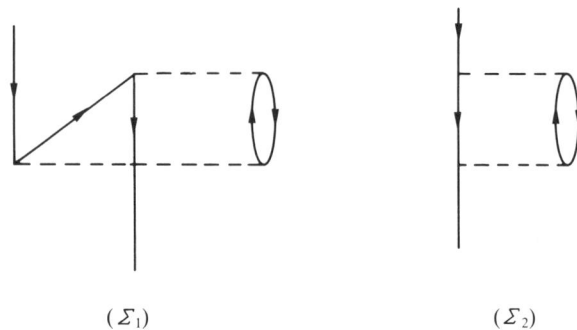

FIG. 5.2. The graphs corresponding to Σ_1 and Σ_2 (Orland and Schaeffer 1978).

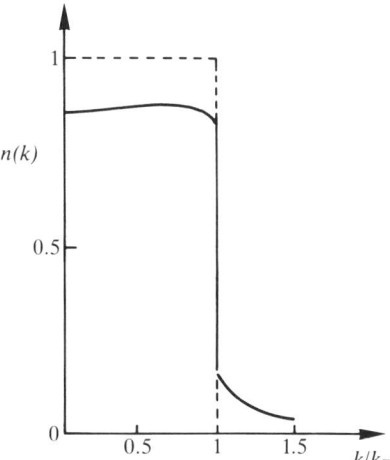

FIG. 5.3. Occupation numbers (Orland and Schaeffer 1978) (solid line) in comparison with the Hartree–Fock result (dashed line).

(4.55) Orland and Schaeffer (1978) obtain for the momentum distribution:

$$n(k) = \theta(k_F - k) - \text{sgn}(x)\frac{\alpha\beta}{2\pi}\frac{k_F^2}{2m}\left[1 + \frac{|x|}{\beta} + \left(2\frac{|x|}{\beta} + \frac{x^2}{\beta^2}\right)\exp\left(\frac{|x|}{\beta}\right)\right.$$
$$\left. \times \text{Ei}\left(-\frac{|x|}{\beta}\right)\right], \quad (5.11)$$

where

$$x = 1 - k^2/k_F^2. \quad (5.12)$$

The result for the momentum distribution in this approximation is compared with the Hartree–Fock result in Fig. 5.3. Estimates are given for the occupation numbers in ^{40}Ca and it is shown that the Hartree–Fock values are reduced for all hole orbitals by about 15%. This means that a very large number of nucleons (about 6) are in the higher shells. A similarly large depletion has also been obtained from more realistic calculations using the Brueckner-matrix and random-phase approximation (RPA) methods (see references in the paper of Orland and Schaeffer (1978)).

A semiclassical model for calculating the second-order mass operator is proposed by Hasse and Schuck (1985a, 1985b). The model is based on the Thomas–Fermi theory for multiparticle-multihole configurations (Ghosh et al. 1983; Blin et al. 1984) and its basic assumption is that the dynamical correlations are produced by 2-particle(hole) 1-hole(particle) intermediate states. The polarization and correlation contributions to the real part of the mass operator (or nucleon–nucleus optical potential) are calculated by Hasse and Schuck (1985b) using the results for the imaginary part (Hasse and Schuck 1985a) and via the subtracted dispersion relations (Luttinger 1961). The semiclassical Hartree–Fock potential (Ring and Schuck 1980) determined by the Gogny D1 effective interaction (Dechargè and Gogny 1980; Gogny and Padjen 1977) and the Perey-Buck (1962) potential is used as the underlying non-local mean-field potential. The semiclassical approximation in the model consists in writing all expressions in Wigner space and then replacing the operators by their classical counterparts and their traces by phase-space integration (Ring and Schuck 1980). Different quantities such as depths, radial dependence, and volume integrals of single-particle potentials, rearrangement energies and effective masses, mean-free paths of nucleons in nuclei, single-particle-level densities, as well as momentum distribution, are calculated by means of the second-order mass operator obtained in the model (Hasse and Schuck 1985b).

The momentum distribution (or occupation probabilities) is expressed by means of the correlation ($\tilde{V}_{corr}$) and polarization ($\tilde{V}_{pol}$) contributions to the real part of the mass operator (Orland and Schaeffer 1978):

$$n(p) = \begin{cases} 1 + \partial \tilde{V}_{pol}/(r, p, E)/\partial E|_{\text{on shell}}, & p < p_F \\ -\partial \tilde{V}_{corr}(r, p, E)/\partial E|_{\text{on shell}}, & p > p_F. \end{cases} \quad (5.13)$$

Using the finite-range force the momentum distribution (Hasse and Schuck 1985b) is calculated by

$$\partial \tilde{V}_{pol;corr}(r, p, E)/\partial E|_{\text{on shell}} = C_b(\hat{U}\hat{V}A_{\partial p, \partial c}), \quad (5.14)$$

where the integral operators $\hat{U}$, $\hat{V}$, and $A_{\partial p, \partial c}$ read:

$$\hat{U} = \frac{k_0}{\sqrt{(2\pi)}R} \int_0^{R_\lambda} dr' \cdot r'[\exp\{-\tfrac{1}{2}k_0^2(r-r')^2\} - \exp\{-\tfrac{1}{2}k_0^2(r+r')^2\}], \quad (5.15)$$

$$\hat{V} = \frac{1}{2p \cdot p_F^4} \int_0^\infty dq \cdot q e^{-2q^2/k_0^2} \int_0^\infty dp_z \theta(k_F(r') - |p_z - \tfrac{1}{2}q|)$$
$$\times [2p_z q\theta(k_F(r') - p_z - \tfrac{1}{2}q) + (k_F^2(r') - (p_z - \tfrac{1}{2}q)^2)$$
$$\times \theta(p_z + \tfrac{1}{2}q - k_F(r'))], \quad (5.16)$$

$$A_{\partial c} = \frac{1}{\pi \lambda} \begin{cases} \dfrac{1}{Z + 2pq - q^2} - \dfrac{1}{p^2 - k_z^2 + Z}, & (p-q)^2 < k_F^2(r) < (p+q)^2 \\ \dfrac{4pq}{4p^2q^2 - (Z - q^2)^2}, & (p+q)^2 < k_F^2(r), \end{cases}$$

$$(5.17)$$

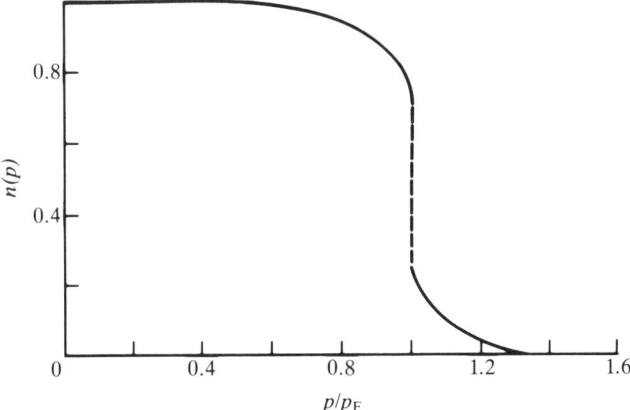

FIG. 5.4. Momentum distribution in the semiclassical model of Hasse and Schuck (1985b).

$$A_{\partial p} = \frac{1}{\pi \lambda} \begin{cases} \dfrac{4pq}{4p^2q^2 - (Z - q^2)^2}, & k_F^2(r) < (p-q)^2 \\ \dfrac{1}{p^2 - k_F^2 - Z} + \dfrac{1}{Z + 2pq + q^2}, & (p-q)^2 < k_F^2(r) < (p+q)^2 \end{cases}$$

(5.18)

In eqns (5.15)–(5.18) R_λ is the classical turning point, $Z = 2p_z qm^*(r)/m^*(r')$; $m^*(r)$ is the effective mass; $k_F(r)$ is the local Fermi-momentum; C_b is a constant; $p_F = 1.495$ fm^{-1}, $\lambda = 46.4$ MeV.

The result for the momentum distribution obtained in the model of Hasse and Schuck (1985b) is shown in Fig. 5.4. As can be seen the correlation and polarization effects smear out the step-function distribution but a finite discontinuity for $p = p_F$ still remains. The value of the gap is 0.5 which is in accordance with the results of Orland and Schaeffer (1978), as well as with those of Flynn *et al.* (1984) and Fantoni and Pandharipande (1984).

It must be noted that using the Gogny force, which is an effective one, means that consideration of this model differs from the treatment of the bare nucleon–nucleon force in perturbation theory.

5.2. A hole-line expansion: the Brueckner–Hartree–Fock approach

The perturbation expansion of the mass operator (5.1) can be analysed in the case when the nucleon–nucleon interaction is relatively weak. In the case of realistic nucleon–nucleon interactions when hard-core components of the forces can exist this expansion is not appropriate for nuclear matter. Brueckner, Bethe (1971) and others (Sprung 1972; Köhler 1975) reordered the perturbation series in order to group together all graphs of the perturbation series which contain the same number of hole lines. This yields the hole-line expansion, or the so-called low-density expansion.

The leading term of the hole-line expansion for the mass operator is obtained by summing all the graphs with a single hole line and has the form:

$$\Sigma_{\text{BHF}}(k; E) = \sum_{j} \theta(k_F - |j|) \langle k, j | g[E + e(j)] | k, j - j, k \rangle, \quad (5.19)$$

where $g(E)$, the reaction matrix, is the solution of the integral (Bethe–Goldstone) equation

$$g(E) = \vartheta + \vartheta \sum_{c,d} \theta(|c| - k_F) \theta(|d| - k_F) \frac{|c, d\rangle \langle c, d|}{E - e(c) - e(d) + i\eta} g(E). \quad (5.20)$$

The expression for Σ_{BHF} (5.19) is illustrated by graphs in Fig. 5.5 where the reaction matrix g is represented by a wiggly line. In eqn (5.20) E is the so-called 'starting energy' and

$$e(d) = d^2/2m + U(d). \quad (5.21)$$

The question of choosing the auxiliary potential $U(d)$ is thoroughly discussed by Mahaux *et al.* (1985) and Jeukenne *et al.* (1976).

The mass operator in the approximation (5.19) is the so-called Brueckner–Hartree–Fock (BHF) approximation.

The formal difference from the Hartree–Fock approximation is that the interaction ϑ in the expression Σ_{1a} (4.49) is replaced by the reaction matrix g which depends on the energy. The BHF method is applicable to arbitrary nucleon–nucleon potentials.

Now we are in a position to find the nucleon momentum distribution using the relation between $n(k)$ and the Green function (4.55).

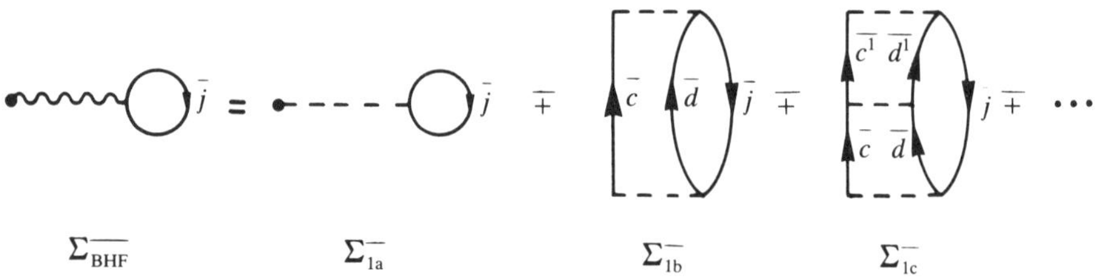

FIG. 5.5. The graph corresponding to Σ_{BHF} (5.19).

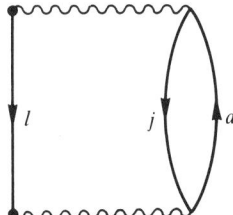

FIG. 5.6. The graph corresponding to Σ_2 (5.24).

The BHF approximation result for the momentum distribution in nuclear matter can be obtained in the form:

$$n_{\text{BHF}}(k) = 1 - \tfrac{1}{2} \sum_{l,c,d} \theta(k_F - |l|)\theta(|c| - k_F)\theta(|d| - k_F)$$
$$\times \frac{|\langle k, l | g[e(k) + e(l)] | c, d - d, c \rangle|^2}{[e(k) + e(l) - e(c) - e(d)]^2}, \quad \text{for } k < k_F, \quad (5.22)$$

and

$$n_{\text{BHF}}(k) = 0, \quad \text{for } k > k_F. \quad (5.23)$$

It is interesting to note that in the BHF approximation there are no particles with momentum $k > k_F$.

The leading term of the $n(k)$ expansion for $k > k_F$ can be obtained from the second term Σ_2 in the hole-line expansion (see Fig. 5.6):

$$\Sigma_2(k, E) = \tfrac{1}{2} \sum_{j,l,a} \theta(k_F - |j|)\theta(k_F - |l|)\theta(|a| - k_F)$$
$$\times \frac{|\langle j, l | g[e(j) + e(l)] | k, a - a, k \rangle|^2}{E + e(a) - e(j) - e(l) - i\eta}. \quad (5.24)$$

In this approximation:

$$n_2(k) = \tfrac{1}{2} \sum_{j,l,a} \theta(k_F - |j|)\theta(k_F - |l|)\theta(|a| - k_F)$$
$$\times \frac{|\langle k, a | g[e(j) + e(l)] | j, l - l, j \rangle|^2}{[e(k) + e(a) - e(j) - e(l)]^2}, \quad \text{for } k > k_F \quad (5.25)$$

In applications of the BHF approximation to finite nuclei it is convenient to present its basic equations in a more familiar form. The BHF energy of the system is written

$$E_{\text{BHF}} = \sum_i \langle i | T | i \rangle + \tfrac{1}{2} \sum_{ij} \langle ig | G | i,j - j,i \rangle, \quad (5.26)$$

where the Brueckner G-matrix is defined by the equation:

$$G = \vartheta + \vartheta \frac{Q}{e} G, \tag{5.27}$$

and

$$e = E(i) + E(j) - E(p) - E(q) \tag{5.28}$$

where i, j are the initial states of the nucleons, and p and q are intermediate states. In eqn (5.27) Q is the Pauli operator equal to 1 if the states p and q are not occupied and zero otherwise.

Self-consistency of the equations is reached when the one-particle energies in (5.28) are given by:

$$E(i) = \langle i | T | i \rangle + \sum_j \langle ij | G(E_i + E_j) | i, j - j, i \rangle. \tag{5.29}$$

The one-particle energies and potentials must also be consistent in the Hartree–Fock sense which means that the following equations exist:

$$T\psi_i(r) + \sum_j \int \psi_j(r') G(r, r'; r_1, r_1') \{\psi_i(r_1)\psi_j(r_1') - \psi_i(r_1')\psi_j(r_1)\}$$
$$\times dr' \, dr_1 \, dr_1' = \varepsilon_i \psi_i(r). \tag{5.30}$$

Because of the difficulties arising from the necessity of attaining double self-consistency, various approximate methods are used in applications of the BHF approximation.

The results for the basic nuclear characteristics such as density distribution, binding energies, and radii are often not satisfactory and depend on the additional approximations used in the method.

Here we present the result of the Brueckner calculations for the ^{16}O nucleon momentum distribution given in the paper of Van Orden *et al.* (1980). The results are illustrated in Fig. 5.7 for the Reid soft-core potential and for the de Tourreil-Sprung super-soft-core potential. In the same figure the results are compared with the momentum distributions obtained from shell-model calculations using harmonic-oscillator and Saxon-Woods potentials. We should mention that this result for the momentum distribution shows a high-momentum tail very far from the shell-model predictions. Remembering the BHF result for nuclear matter ($n_{\text{BHF}} = 0$ at $k > k_F$) this may be due to the rather crude approximations used in the calculations.

Algebraic expressions for the momentum distribution in the correlated ground state of a hard-sphere dilute Fermi-gas are obtained by Sartor and Mahaux (1980a). The calculation is carried out using the low-density expansion of the mass operator up to terms of order $(k_F c)^2$, where k_F is

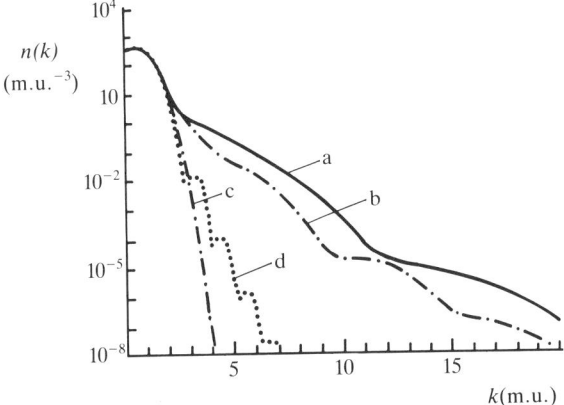

FIG. 5.7. Nucleon momentum distribution for ^{16}O (Van Orden *et al.* 1980) ($mu = m_\pi c = 139.57$ MeV/c). curve a): correlated distribution using the Reid potential; curve b): correlated distribution using the Sprung potential; curve c): shell-model harmonic-oscillator momentum distribution; curve d): shell-model Saxon–Woods momentum distribution.

the Fermi-momentum and c is the hard-core radius. The momentum distribution $n(k)$ of a hard-sphere dilute Fermi-gas of nucleons (with spin-isospin degeneracy $v = 4$) at $k_F c = 0.7$ is presented in Fig. 5.8. It is pointed out that the shape of $n(k)$ is independent of the choice of v and $k_F c$. It is shown (Sartor and Mahaux 1980a) that this momentum

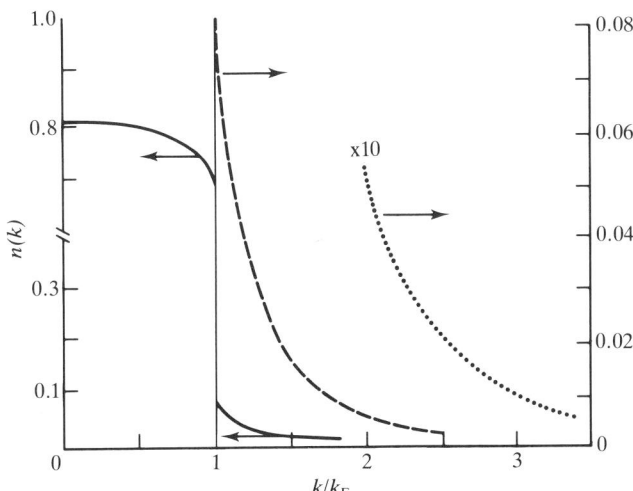

FIG. 5.8. Momentum distribution of a hard-sphere dilute Fermi-gas for the parameter $v = 4$, $k_F c = 0.7$ calculated by Sartor and Mahaux (1980a).

distribution is in agreement with the one obtained by Czyż and Gottfried (1961) and differs from the result of Belyakov (1961) in the region of momenta $1 < k/k_F < 3$. The behaviour of $n(k)$ at large momenta is found to be $k^{-(4+m)}$, where the value of m depends on the asymptotic behaviour of the imaginary part of the mass operator.

5.3. The exp(S) approach

In the coupled-cluster many-body (the so-called exp(S)) theory developed by Kümmel (1971), Kümmel and Lührmann (1972), Lührmann and Kümmel (1973), Zabolitzky (1973, 1974a, 1974b), Gari et al. (1976), Kümmel and Zabolitzky (1973), Emrich et al. (1977), Kümmel et al. (1978), Zabolitzky and Ey (1978) the ground-state wavefunction for a doubly closed shell nucleus is constructed in the form:

$$|\psi\rangle = \exp(S)|\Phi\rangle. \quad (5.31)$$

In this ansatz $|\Phi\rangle$ is the determinant of one-particle orbitals, S is the sum of particle-hole excitation amplitudes

$$S = \sum_{n=1}^{A} S_n, \quad (5.32)$$

and

$$S_n = \frac{1}{(n!)^2} \sum_{\nu_i, \rho_i} \langle \rho_1, \ldots \rho_n | S_n | \nu_1, \ldots \nu_n \rangle_A a^+_{\rho_1} \ldots a^+_{\rho_n} a_{\nu_n} \ldots a_{\nu_1}. \quad (5.33)$$

Here ν, μ, $\lambda(\rho, \sigma, \tau)$ label normally occupied (unoccupied) states and α, β, γ are used for both kinds of states. A is the number of nucleons in the system. The subscript 'A' implies antisymmetrization:

$$|\nu_1 \nu_2\rangle_A = |\nu_1 \nu_2\rangle - |\nu_2 \nu_1\rangle. \quad (5.34)$$

The operator S_n produces linked n-particle n-hole excitations.

The ground-state wavefunction $|\psi\rangle$ satisfies the Schrödinger equation with a given Hamiltonian $\hat{H} = \hat{T} + \hat{V}$ from which a hierarchy of non-linear coupled equations for the amplitude $\langle \ldots | S_n | \ldots \rangle$ and the single-particle orbitals can be obtained.

The Schrödinger equation

$$\hat{H}|\psi\rangle = E|\psi\rangle \quad (5.35)$$

can be written in the form:

$$(e^{-S}\hat{H}e^{S} - E)|\Phi\rangle = 0. \quad (5.36)$$

Then projecting on $\langle\Phi|$, $\langle\Phi| a^+_\rho a_\nu$, etc. gives coupled equations for E, $\langle\rho|S_1|\nu\rangle$, etc. A general solution of these equations is not possible in practice, but one can hope that after a suitable truncation the remaining final set of coupled equations will be tractable. The truncation of the system of equations neglecting subsequently all terms S_n ($n \geq 2$), S_n ($n \geq 3$), S_n ($n \geq 4$) etc. leads to the generalized Hartree–Fock equation, generalized BHF equation, generalized three-particle Bethe-Faddeev equation and so on. For example the generalized BHF equations can be written as follows:

$$T_{x_1}\langle x_1| \psi_1 |v_1\rangle + \langle x_1 | U |v_1\rangle + \sum_{v}^{A} \langle x_1 v| TS_2 |v_1 v\rangle = \sum_{v} h_{vv_1}\langle x_1| \psi_1 |v\rangle, \quad (5.37)$$

where
$$\langle x_1| \psi_1 |v_1\rangle = \langle x_1|v_1\rangle + \langle x_1| S_1 |v_1\rangle \quad (5.38)$$

represent 'single-particle states.' Here $\langle x_1| S_1 |v_1\rangle = \sum_\rho \langle x_1|\rho\rangle \langle\rho| S_1 |v_1\rangle$ is the coordinate-space representation of S_1, and $\langle x | v\rangle$ and $\langle x | \rho\rangle$ are occupied and intermediate oscillator states.

In eqn (5.37)

$$\langle x| U |v_1\rangle = \sum_{v}^{A} \langle xv| V\psi_2 |v_1 v\rangle(1 - d_v) \quad (5.39)$$

is the (renormalized) single-particle potential, and

$$d_v = \sum_{v'} \int dx_1 \, dx_2 \langle x_1 x_2| S_2 |vv'\rangle^* \langle x_1 x_2| S_2 |vv'\rangle \quad (5.40)$$

is the 'depletion factor' and

$$h_{vv'} = \sum_{v''} \langle vv''| V\psi_2 |v'v''\rangle(1 - d_{v''}) + \langle v| T\psi_1 |v'\rangle \quad (5.41)$$

is the (renormalized) BHF matrix. The abbreviation

$$\langle xx'| V\psi_2 |vv'\rangle = \int dx'' \, dx''' \langle xx'| V |x''x'''\rangle \langle x''x'''| \psi_2 |vv'\rangle \quad (5.42)$$

is used, where

$$\langle x_1 x_2| \psi_2 |v_1 v_2\rangle = \mathcal{A} \langle x_1| \psi_1 |v_1\rangle \langle x_2| \psi_1 |v_2\rangle + \langle x_1 x_2| S_2 |v_1 v_2\rangle \quad (5.43)$$

is the so-called Bethe–Goldstone pair wavefunction.

Now the one-body density matrix D_1 can be introduced by means of

$$\langle \alpha_1| D_1 |\alpha_2\rangle = \langle \psi| a^+_{\alpha_2} a_{\alpha_1} |\psi\rangle / \langle \psi | \psi\rangle, \quad (5.44)$$

which can be used for the determination of the average value of an arbitrary one-particle operator. The diagonal element of the one-body density matrix in the q representation is the nucleon momentum distribution which can be written in the form (Zabolitzky and Ey 1978):

$$n(q) = \sum_{\alpha_1 \alpha_2} \langle q | \alpha_1 \rangle \langle \alpha_1 | D_1 | \alpha_2 \rangle \langle \alpha_2 | q \rangle, \qquad (5.45)$$

where $\langle q | \alpha \rangle$ is the complete orthonormal set of orbitals in the momentum space.

The one-body density matrix for closed-shell nuclei can be obtained in the accepted approximation ($S_n = 0$, $n \geq 3$, $\langle S_1 \rangle = 0$):

$$\begin{aligned}
\langle \alpha_1 | D_1 | \alpha_2 \rangle &= \sum_v \langle \alpha_1 | v \rangle \langle v | \alpha_2 \rangle \\
&+ \sum_{v_1 v_2 \rho_1 \rho'_1 \rho_2} \langle \alpha_1 | \rho_1 \rangle \langle v_1 v_2 | S_2^+ | \rho_1 \rho_2 \rangle \langle \rho'_1 \rho_2 | S_2 | v_1 v_2 \rangle \\
&\quad \times \langle \rho'_1 | \alpha_2 \rangle / (1 + 2K) \\
&- \sum_{v_1 v'_1 v_2 \rho_1 \rho_2} \langle \alpha_1 | v_1 \rangle \langle v_1 v_2 | S_2^+ | \rho_1 \rho_2 \rangle \langle \rho_1 \rho_2 | S_2 | v'_1 v_2 \rangle \\
&\quad \times \langle v'_1 | \alpha_2 \rangle / (1 + 2K), \qquad (5.46)
\end{aligned}$$

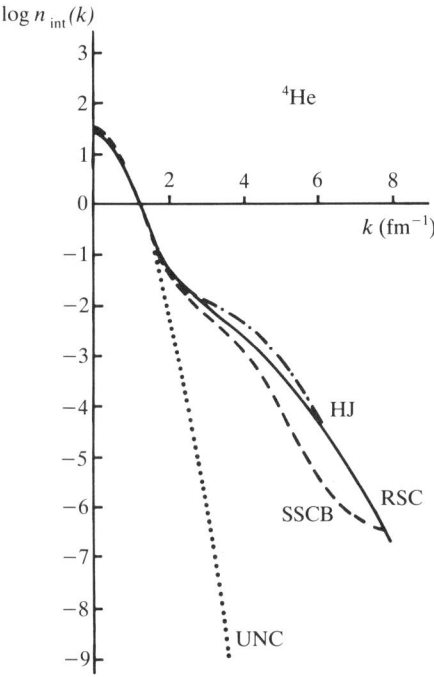

FIG. 5.9. Nucleon momentum distribution for ^{4}He (Zabolitzky and Ey 1978). HJ: Hamada–Johnston potential; RSC: Reid soft-core potential; SSCB: de Tourreil–Sprung super-soft-core potential B; UNC: uncorrelated, for the RSC potential.

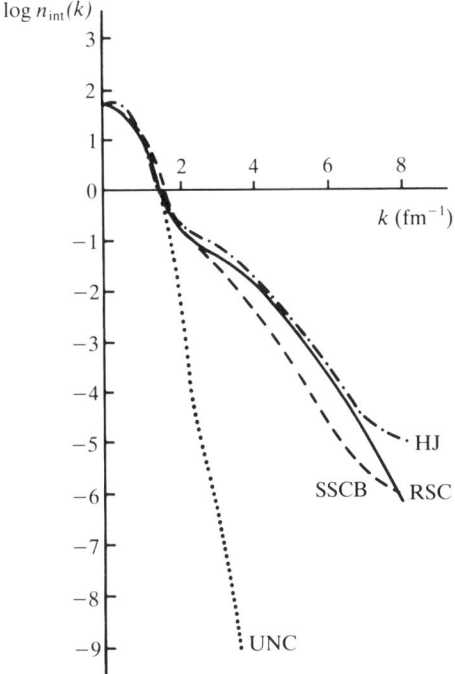

FIG. 5.10 Same as Fig. 5.9 for ^{16}O (Zabolitzky and Ey 1978).

with

$$K = \frac{1}{A} \sum_{\substack{v_1 v_2 \\ \rho_1 \rho_2}} \langle v_1 v_2 | S_2^+ | \rho_1 \rho_2 \rangle \langle \rho_1 \rho_2 | S_2 | v_1 v_2 \rangle_A. \quad (5.47)$$

Then the momentum distribution can be decomposed in a schematic way:

$$n(q) = n_1(q) + n_2(q) + n_3(q), \quad (5.48)$$

where each term corresponds to a term in eqn (5.46). The first term $n_1(q)$ describes the momentum distribution due to the uncorrelated motion of particles in occupied orbitals; these are Brueckner or maximum-overlap orbitals obtained by the condition

$$\langle \psi | \Phi \rangle = \max_{\langle x | v \rangle}. \quad (5.49)$$

The remaining two terms $n_2(q)$ and $n_3(q)$ originate from the population of normally unoccupied orbitals, $n_2(q)$, and from the corresponding depletion of normally occupied orbitals, $n_3(q)$. They are consequences of the correlations contained in those parts of the full wavefunction ψ which cannot be expressed in terms of a single Slater determinant.

Numerical results for $n(q)$ in ^{4}He and ^{16}O obtained using different nucleon–nucleon potentials: Hamada-Johnston, Reid soft-core (RSC) and the de Tourreil-Sprung super-soft-core potentials are presented in Figs. 5.9 and 5.10 (Zabolitzky and Ey 1978) and compared with the uncorrelated momentum distribution calculated with the RSC potential.

The strong deviation of $n(q)$ from the uncorrelated distribution at $q > 2 \text{ fm}^{-1}$ corresponds to the account of nucleon–nucleon correlations in the approach.

The local density (or the form factor) can also be calculated by this method (Gari *et al.* 1976) using the one-body density matrix (5.44), in which correlations are taken into account. It turns out that there is no significant influence of correlations in the form factor.

So the conclusion can be drawn that the nucleon momentum distribution is quite sensitive to the nucleon–nucleon correlations especially at large q ($q \geqslant 2 \text{ fm}^{-1}$) which is not the case for the form factor. These peculiarities of the two bulk characteristics $n(q)$ and $\rho(r)$ must be taken into account in experiments to determine them.

6

PHENOMENOLOGICAL CORRELATION METHODS

In this chapter, Jastrow-type methods are briefly discussed. Variational Jastrow-type calculations of the momentum distributions for nuclear matter and for nuclei with $A = 3$ and $A = 4$ are given in Section 6.2. A simple phenomenological model for introducing dynamical short-range and tensor nucleon–nucleon correlations is presented in Section 6.3. Their effects on the momentum distributions and form factors are studied in the same section.

6.1. Jastrow-type methods

In order to take into account the correlations at small distances due to the hard-core nucleon–nucleon potential, which are not taken into account in the determinant form of the wavefunction of the system, a method was proposed by Jastrow (1955). This method, known as the Jastrow-correlation method was widely used with different modifications in many fields of physics concerned with Fermi systems.

The starting point of the method is the ansatz for the wavefunction of A fermions

$$\psi(r_1, \ldots r_A) = \frac{1}{\sqrt{C_A}} \prod_{1 \leq i < j \leq A} f(r_{ij}) \Phi(r_1, \ldots r_A), \tag{6.1}$$

where the normalization constant C_A is determined by

$$\int |\psi(r_1, \ldots r_A)|^2 \, dr_1 \ldots dr_A = 1. \tag{6.2}$$

The function Φ in (6.1) is a Slater determinant constructed from A single-particle wavefunctions $\varphi_\alpha(r)$

$$\Phi(r_1, \ldots r_A) = \det|\varphi_\alpha(r_i)|, \tag{6.3}$$

such that $\varphi_\alpha(r)$ correspond to the lowest bounded A orbitals.

The correlation function $f(r_{ij}) \equiv f(|r_i - r_j|)$ depends on the absolute value $|r_i - r_j|$ so that the construction (6.1) is an antisymmetrized and normalized wavefunction. The function f is chosen in such a way that the wavefunction ψ vanishes when the distance between particles $|r_i - r_j|$ becomes smaller than the radius of the nucleon–nucleon repulsive core. In general it satisfies the conditions

$$f(r_{ij}) = 0, \quad |r_i - r_j| \leq r_c,$$
$$f(r_{ij}) = 1, \quad |r_i - r_j| \to \infty. \qquad (6.4)$$

The wavefunction ψ (6.1) is used as a trial function for a variational determination of the system's energy with a given Hamiltonian. In the case of Hamiltonian $\hat{H}$ with two-nucleon forces $\vartheta(r_{ij})$ the expectation value of $\hat{H}$

$$\langle \psi | \hat{H} | \psi \rangle = \langle \psi | \hat{T} | \psi \rangle + \langle \psi | \hat{V} | \psi \rangle, \qquad (6.5)$$

contains the total kinetic energy:

$$\langle \psi | \hat{T} | \psi \rangle = -\frac{\hbar^2}{2m} \int \nabla_{r_1} \cdot \nabla_{r_1'} \rho(r_1, r_1')\big|_{r_1 = r_1'} dr_1, \qquad (6.6)$$

and the total potential energy (for a Wigner force):

$$\langle \psi | \hat{V} | \psi \rangle = \frac{1}{2} \int dr_1 \, dr_2 \vartheta(r_{12}) \rho(r_1, r_2; r_1', r_2')\big|_{\substack{r_1 = r_1' \\ r_2 = r_2'}}. \qquad (6.7)$$

The one- and two-body density matrices are defined as:

$$\rho(r, r') = A \int \psi^*(r, r_2, \ldots r_A) \psi(r', r_2, \ldots r_A) \, dr_2 \ldots dr_A, \qquad (6.8)$$

$$\rho(r_1, r_2; r_1', r_2') = \tfrac{1}{2} A(A-1) \int \psi^*(r_1, r_2, r_3, \ldots r_A)$$
$$\times \psi(r_1', r_2', r_3, \ldots r_A) \, dr_3 \ldots dr_A. \qquad (6.9)$$

The variations of the expectation value of $\hat{H}$ with respect to single-particle wavefunctions $\varphi_\alpha(r)$ and the correlation function $f(r_{ij})$ lead, in principle, to the corresponding Euler–Lagrange equations. The difficulties of solving these equations makes it necessary to introduce some approximations and to develop appropriate techniques to treat the problem. These difficulties, because of the form of the wavefunction ψ, are related to the impossibility of writing explicitly the expression for the energy and even for the normalization constant. We now discuss some of these approximations.

The perturbation expansion method (Gaudin et al. 1971) is developed for calculations of the one- and two-body density matrices.

The one-body density matrix $\rho(r_1, r_1')$ (6.8) is written as an expansion

in terms of the functions
$$g(r) = |f(r)|^2 - 1, \qquad (6.10)$$
$$h(r) = f(r) - 1, \qquad (6.11)$$

$$\rho(r_1, r_1') = \frac{A}{C_A} \left\{ \Delta_1 \binom{r_1'}{r_1} + \frac{1}{1!} \int W_1(r_2) \Delta_2 \binom{r_1' \ r_2}{r_1 \ r_2} dr_2 \right.$$
$$+ \frac{1}{2!} \int W_2(r_2, r_3) \Delta_3 \binom{r_1' \ r_2 \ r_3}{r_1 \ r_2 \ r_3} dr_2 \, dr_3$$
$$\left. + \ldots + \frac{1}{(A-1)!} \int W_{A-1} \Delta_A \, dr_2 \ldots dr_A \right\}, \qquad (6.12)$$

where for example:
$$W_1(r_i) = h^*(r_1 - r_i) + h(r_1' - r_i) + h^*(r_1 - r_i)h(r_1' - r_i),$$
$$W_2(r_i, r_j) = g(r_{ij}) + h^*(r_1 - r_j)h(r_1' - r_j) \qquad (6.13)$$
$$+ h^*(r_1 - r_j)h(r_1' - r_j) + h^*(r_1 - r_j)g(r_{ij}) + \ldots$$

$$\Delta_p \binom{r_1' \ldots r_p'}{r_1 \ldots r_p} = \det |\tilde{\rho}(r_i, r_j')|, \qquad i,j \in [1, p], \qquad (6.14)$$

$$\tilde{\rho}(r_i, r_j) = \sum_{\alpha \in F} \varphi_\alpha^*(r_i) \varphi_\alpha(r_j), \qquad (6.15)$$

$$\det |\tilde{\rho}(r_i, r_j)| = |\Phi(r_1, \ldots r_A)|^2, \qquad i,j \in [1, A], \qquad (6.16)$$

$$C_A = A! \left\{ 1 + \sum_{p=2}^{A} \left[\frac{1}{p!} \int W_p(r_1, \ldots r_p) \Delta_p(r_1, \ldots r_p) \, dr_1 \ldots dr_p \right] \right\}, \qquad (6.17)$$

$$\Delta_p(r_1, r_2, \ldots r_p) = \det |\tilde{\rho}(r_i, r_j)|, \qquad i,j \in [1, p]. \qquad (6.18)$$

A similar expansion of the two-body density matrix is also obtained by Gaudin *et al.* 1971.

The one-body density matrix (6.8) makes it possible to calculate the expectation values of the one-body operators corresponding to observable quantities.

In the case of nuclear matter one can calculate the nucleon momentum distribution using the expansion of $\rho(r, r')$ (6.12), which in the lowest-order-cluster (LOC) approximation in g is given by (Flynn *et al.* 1984).

$$\rho^{(1)}(r_1, r_1') \simeq \rho_{LOC}^{(1)}(r_1, r_1') \equiv \tilde{\rho}(r_1, r_1')$$
$$+ \int dr_2 [f(|r_1 - r_2|) f(|r_1' - r_2|) - 1]$$
$$\times [4\tilde{\rho}(r_2, r_2)\tilde{\rho}(r_1, r_1') - \tilde{\rho}(r_1, r_2)\tilde{\rho}(r_2, r_1')]$$
$$- \iint dr_2 \, dr_3 [f^2(r_{23}) - 1] \tilde{\rho}(r_1, r_2)$$
$$\times [4\tilde{\rho}(r_2 r_1')\tilde{\rho}(r_3, r_3) - \tilde{\rho}(r_2, r_3)\tilde{\rho}(r_3, r_1')], \qquad (6.19)$$

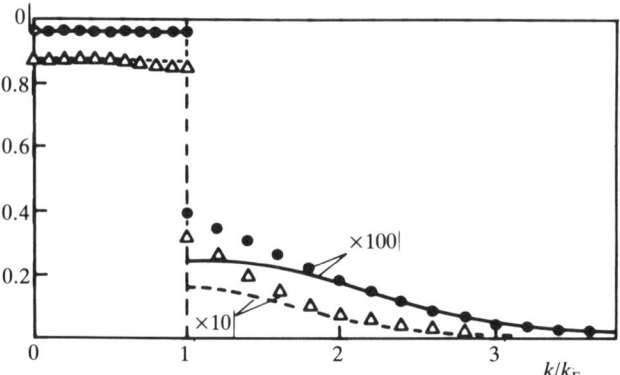

FIG. 6.1. Momentum distribution of symmetrical nuclear matter at $k_F = 1.32\,\text{fm}^{-1}$ (Flynn et al. 1984). Solid curve: FHNC/O with $\kappa_{\text{dir}} = 0.063$. Circular data points: LOC with $\kappa_{\text{dir}} = 0.063$. Dashed curve: FHNC/O with $\kappa_{\text{dir}} = 0.230$. Triangle data points: LOC with $\kappa_{\text{dir}} = 0.230$.

where $\tilde{\rho}(r, r')$ (see eqn (6.15)) is the density matrix corresponding to the wavefunction $\Phi(f \equiv 1)$.

The momentum distribution in the LOC approximation is calculated for symmetrical nuclear matter (Flynn et al. 1984) and the results are illustrated in Fig. 6.1.

The calculations are carried out at two values of the Jastrow wound parameter

$$\kappa_{\text{dir}} = \frac{2}{3\pi^2} \int [f(x) - 1]^2 \, dx \tag{6.20}$$

using the Gaussian form for the correlation function

$$f(r) = 1 - \exp(-\beta^2 r^2), \tag{6.21}$$

with $\beta = 1.69\,\text{fm}^{-1}$ for $\kappa_{\text{dir}} = 0.063$ and $\beta = 1.1$ for $\kappa_{\text{dir}} = 0.230$.

In Fig. 6.1 are the results of calculations using the leading approximation to $n(k)$ within the irreducible cluster formalism (Ristig and Clark 1976). It is shown by Ristig and Clark (1976) that the cluster expansion of $n(k)$ can be partially resummed which leads to the expression of $n(k)$:

$$\begin{aligned} n(k) &= n[\theta(k_F - k)M(k) + N(k)], \\ n &= \exp Q(0). \end{aligned} \tag{6.22}$$

in terms of quantities $Q(0)$, $N(k)$, and $M(k)$ which have irreducible structures. The so-called Fermi-hypernetted-chain (FHNC) technique (see, for example Fantoni (1978); Fantoni and Rosati (1975)) is used in this approximation for calculations of $n(k)$.

The presented formalism is applicable not only for infinite systems but also to finite nuclei. For this purpose the one-particle wavefunctions $\varphi_\alpha(r)$ in (6.3) are chosen in the form of harmonic-oscillator wavefunctions, of the Saxon–Woods type or as the solutions of the self-consistent Hartree–Fock equations.

Applications of the present correlation method in calculations of the nucleon momentum distribution and the form factor (or density distribution) in the case of ^{4}He have already been done (Bohigas and Stringari 1980; Dal Rì et al. 1982).

Using LOC, harmonic-oscillator single-particle wavefunctions and the correlation factor $f(r)$ in the form:

$$f(r) = 1 - \exp(-\beta^2 r^2), \tag{6.23}$$

analytical expressions for the normalized one-body density matrix $\rho(r_1, r_2)$, the form factor $\mathcal{F}(q) = \int e^{i q \cdot r} \rho(r)\, dr$ and nucleon momentum distribution are obtained:

$$\rho(r_1, r_2) = \frac{\alpha^3}{\pi^{\frac{3}{2}}} \left\{ \delta \exp[-\tfrac{1}{2}\alpha^2(r_1^2 + r_2^2)] - \frac{3}{(1+y)^{\frac{3}{2}}} \right.$$
$$\times \left\{ \exp\left[-\frac{\alpha^2}{2}\left(\frac{1+3y}{1+y}\right)(r_1^2 + r_2^2)\right] + (r_1 \leftrightarrow r_2) \right\} + \frac{3}{(1+2y)^{\frac{3}{2}}}$$
$$\left. \times \exp\left[-\tfrac{1}{2}\alpha^2\left((1+2y)(r_1^2 + r_2^2) - \frac{2y^2}{1+2y}(r_1 + r_2)^2\right)\right] \right\}, \tag{6.24}$$

$$\mathcal{F}(q) = \delta \exp\left(-\frac{q^2}{4\alpha^2}\right) - 6(1+2y)^{\frac{3}{2}} \exp\left(-\frac{1+y}{1+2y}\frac{q^2}{4\alpha^2}\right)$$
$$+ 3(1+4y)^{-\frac{3}{2}} \exp\left(-\frac{1+2y}{1+4y}\frac{q^2}{4\alpha^2}\right), \tag{6.25}$$

$$n(k) = \frac{1}{\alpha^3 \pi^{\frac{3}{2}}} \left\{ \delta \exp\left(-\frac{k^2}{\alpha^2}\right) - 6(1+3y)^{-\frac{3}{2}} \exp\left(-\frac{1+2y}{1+3y}\frac{k^2}{\alpha^2}\right) \right.$$
$$\left. + 3\left[(1+2y)(1+4y)^{-\frac{3}{2}} \exp\left(-\frac{1}{1+2y}\frac{k^2}{\alpha^2}\right)\right] \right\}, \tag{6.26}$$

where α is the harmonic-oscillator parameter,

$$y^{-1} = \alpha^2/\beta^2, \qquad \delta = 1 + 6(1+2y)^{-\frac{3}{2}} - 3(1+4y)^{-\frac{3}{2}}.$$

The calculations are carried out at the following values of parameters: $\alpha = 0.82$ fm^{-1}, $\beta = 1.40$ fm^{-1}. The numerical results for $\mathcal{F}(q)$ and $n(k)$ are presented in Figs. 6.2 and 6.3.

As can be seen the form factor is in good agreement with the experimental data. The important feature of the momentum distribution

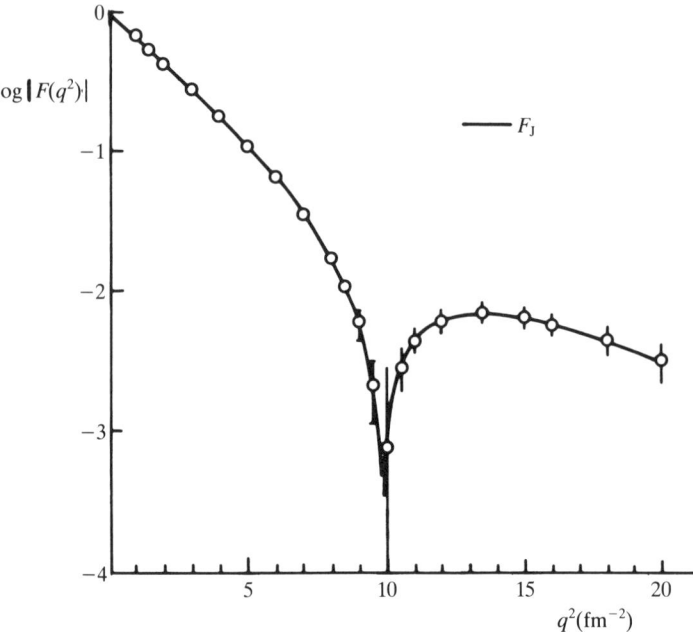

FIG. 6.2. Form factor of ^{4}He. Solid line: Jastrow form factor (Bohigas and Stringari 1980).

is the long tail beyond $k = 2 \text{ fm}^{-1}$ which is in accordance with the $\exp(S)$ result of Zabolitzky and Ey (1978). This result shows the role of the Jastrow-type short-range correlations on the high-momentum components of the momentum distribution.

The envelope of $n(k)$ in Fig. 6.3 is calculated on the basis of single-determinant wavefunctions fitted to reproduce in the best way the form factor $\mathscr{F}(q)$ for ^{4}He. We can see there is a discrepancy between this and the Jastrow momentum distribution (at k larger than 2 fm^{-1}). The result shows that an independent-particle wavefunction (Slater determinant) cannot reproduce simultaneously the form factor and the momentum distribution of a correlated system. It must be noted, however, that the discrepancy between the two momentum distributions is significantly smaller than the one obtained by Zabolitzky and Ey (1978). The important conclusion made by Bohigas and Stringari (1980) is that the presence of short-range correlations may be inferred by investigating simultaneously both quantities: the form factor and the momentum distribution.

The exact Jastrow calculations for $\mathscr{F}(q)$ and $n(k)$ using the correlation

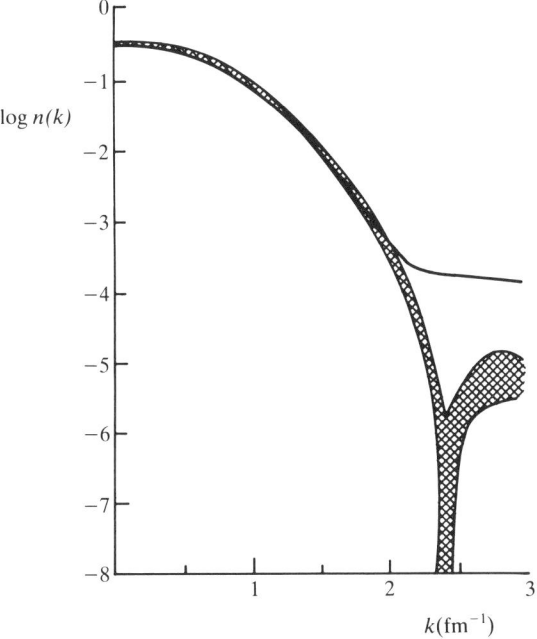

FIG. 6.3. Nucleon momentum distribution of ^{4}He. Solid line: Jastrow momentum distribution $n_J(k)$; dashed area: envelope of $n_{SD}(k)$ corresponding to Slater determinant calculations (Bohigas and Stringari 1980).

factor in the form

$$f(|\mathbf{r}_i - \mathbf{r}_j|) = 1 - c_1 e^{-\beta_1^2 r_{ij}^2} - c_2 e^{-\beta_2^2 r_{ij}^2}, \tag{6.27}$$

and harmonic-oscillator single-particle wavefunctions are carried out and compared with LOC-approximation results in the paper of Dal Rì et al. (1982) and with harmonic-oscillator shell-model calculations. The results are illustrated in Figs. 6.4 and 6.5.

The comparison between exact Jastrow calculations and those from the LOC approximation (in the case of ^{4}He) indicates that the latter works rather well for a quantitative analysis. In particular, it accounts for the high-momentum components in the nucleon momentum distribution.

Interesting calculations of the density distribution and the charge form factor in ^{40}Ca have been carried out by Gaudin et al. (1971) in the LOC approximation of the Jastrow-correlation method. Their results for the density distributions are given in Fig. 6.6. It is remarkable that the Jastrow result (solid line) for $\rho(r)$ is very close to the calculated curve (dashed line) using a Slater-determinant function constructed from natural orbitals. The latter are obtained from diagonalization of the

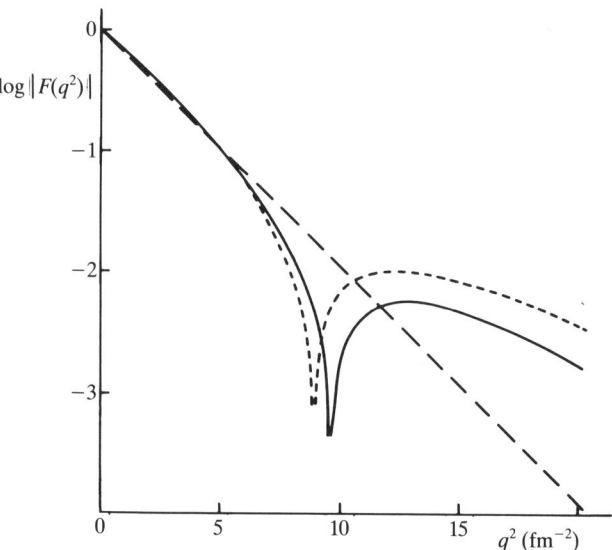

FIG. 6.4. Elastic form factor of ^{4}He (Dal Rì *et al.* 1982). Solid line: exact Jastrow calculations; short dash: LOC approximation; long dash: harmonic-oscillator model calculation.

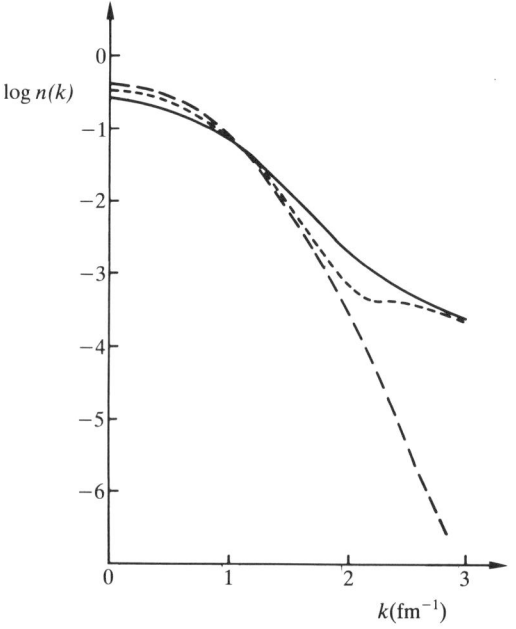

FIG. 6.5. Nucleon momentum distribution of ^{4}He (Dal Rì *et al.* 1982). solid line: exact Jastrow calculation; short dash: LOC approximation; long dash: harmonic-oscillator shell-model calculation.

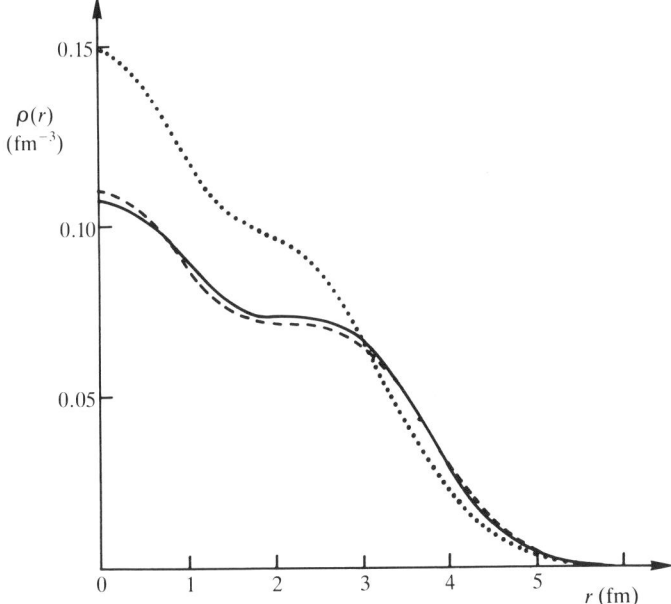

FIG. 6.6. Charge density of ^{40}Ca (not corrected for finite size of the proton) (Gaudin et al. 1971). Solid line: Jastrow method; dashed line: uncorrelated density calculated with natural orbitals obtained by diagonalizing the density matrix of the Jastrow wavefunction; dotted line: uncorrelated density calculated with harmonic-oscillator single-particle wavefunctions ($\alpha = 0.55$ fm^{-1}).

Jastrow one-body density matrix. The dotted curve represents the calculations with a Slater determinant Φ built up from harmonic-oscillator functions which enters in the total Jastrow function ψ. This result raises the question as to whether the analysis usually made of elastic electron scattering can provide a measure of the nucleon–nucleon correlations. It could be very interesting making a similar analysis for the nucleon momentum distribution.

6.2. Variational Jastrow-type calculations

In the paper of Fantoni and Pandharipande (1984) the nucleon momentum distribution $n(k)$ in nuclear matter has been calculated using a realistic Hamiltonian. The nuclear-matter wavefunction is approximated by the variational wavefunction containing: central, spin, isospin, tensor, and spin-orbit pair correlations:

$$\psi(x_1, \ldots x_A) = \left(S \prod_{i>j=1}^{A} \mathcal{F}_{ij} \right) \Phi(x_1, \ldots x_A), \tag{6.28}$$

where $\mathcal{F}_{ij}$ are pair-correlation operators:

$$\mathcal{F}_{ij} = \sum_p f^p(r_{ij}) O^p_{ij}, \tag{6.29}$$

$$O^{p=1,\ldots 8}_{ij} = 1, (\boldsymbol{\tau}_i \cdot \boldsymbol{\tau}_j), (\boldsymbol{\sigma}_i \cdot \boldsymbol{\sigma}_j), (\boldsymbol{\sigma}_i \cdot \boldsymbol{\sigma}_j)(\boldsymbol{\tau}_i \cdot \boldsymbol{\tau}_j), S_{ij},$$
$$S_{ij}(\boldsymbol{\tau}_i \cdot \boldsymbol{\tau}_j), \boldsymbol{L} \cdot \boldsymbol{S}, \boldsymbol{L} \cdot \boldsymbol{S}(\boldsymbol{\tau}_i \cdot \boldsymbol{\tau}_j). \tag{6.30}$$

S is the symmetrizer, and $\Phi(x_1, \ldots x_A)$ is a determinant built up from plane waves. In eqn (6.28) x_α denotes the position r_α and the spin and isospin of particle α. The $f^p(r_{ij})$ are parametrized in a convenient way and the values of the parameters are determined by minimizing the energy expectation value. The variational calculation of the momentum distribution $n(k)$ is carried out using the one-body density matrix $\rho(r_1, r'_1) = \rho(|r_1 - r'_1|)$:

$$\rho(|r_1 - r'_1|) = \tfrac{1}{4} A \frac{\int dr_2 \ldots dr_A \psi^+(x_1, x_2, \ldots x_A)\psi(x'_1, x_2, \ldots x_A)}{\int dr_1 \ldots dr_A |\psi|^2}. \tag{6.31}$$

The factor $\tfrac{1}{4}$ in (6.31) takes into account the spin-isospin degeneracy.

The power-series method (Fantoni and Rosati 1974) is applied to expand

$$n(k) = \int dr e^{-i\boldsymbol{k}\cdot\boldsymbol{r}} \rho(|r_1 - r'_1|), \tag{6.32}$$

where $\rho(|r_1 - r'_1|)$ in eqn (6.31) is obtained by the variational wavefunction reported by Lagaris and Pandharipande (1981).

Since the results of the variational calculations are not satisfactory at $k \sim k_F$ calculations of $n(k)$ are carried out with the wavefunction generalized by using perturbation theory in a correlated basis (CBF) and it now contains the variational and second-order CBF terms:

$$\psi_{\text{CBF}} = \psi + \tfrac{1}{4} \sum_{\substack{h_1 h_2 \\ p_1 p_2}} \alpha(h_1 h_2 p_1 p_2) |h_1 h_2 p_1 p_2\rangle, \tag{6.33}$$

where

$$\alpha(h_1 h_2 p_1 p_2) = \frac{\langle h_1 h_2 p_1 p_2 | H - E^v | \psi \rangle}{e^v(h_1) + e^v(h_2) - e^v(p_1) - e^v(p_2)}. \tag{6.34}$$

$e^v(k)$ are variational single-particle energies, and E^v is the variational ground-state energy. The sum in eqn (6.33) is over $2p-2h$ states. Up to terms of order α^2 the momentum distribution is obtained in the form

$$n^{(2)}_{\text{CBF}}(k) = n(k) + \delta n^{(2)}(k), \tag{6.35}$$

where $n(k)$ is the variational momentum distribution and $\delta n^{(2)}(k)$ is the second-order ($\sim \alpha^2$) CBF correction (Fantoni and Pandharipande 1984).

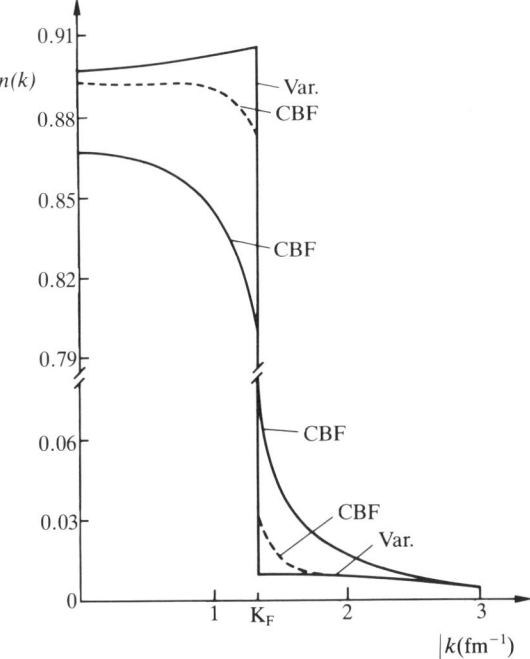

FIG. 6.7. The occupation probabilities in equilibrium nuclear matter obtained from variational (var.) and CBF calculations (Fantoni and Pandharipande 1984). The full and broken lines (CBF) show results of second-order CBF calculations with and without tensor operators, respectively.

The nucleon momentum distribution in equilibrium nuclear matter obtained from variational and CBF calculations is presented in Fig. 6.7.

The full and broken lines show results of second-order CBF calculations with and without tensor operators, respectively. The most interesting result here is the important role of tensor correlation terms in ψ. It is seen that the second-order CBF corrections are reduced by almost a factor of four when the tensor components (S_{ij}) in ψ are not included in the calculations. The role of the tensor correlations underlined here is obviously very important and must be carefully examined in finite nuclear systems.

In the paper of Schiavilla et al. (1986) a general Monte Carlo method of studying the momentum distributions of nucleons and nucleon clusters in nuclei is developed. The method is used to calculate the momentum distributions of protons and neutrons in nuclei with $A = 3$ and $A = 4$ with improved variational wavefunctions. The calculations are based on a realistic Hamiltonian that includes the three-nucleon interaction, and give reasonable binding energies and densities for light nuclei and nuclear

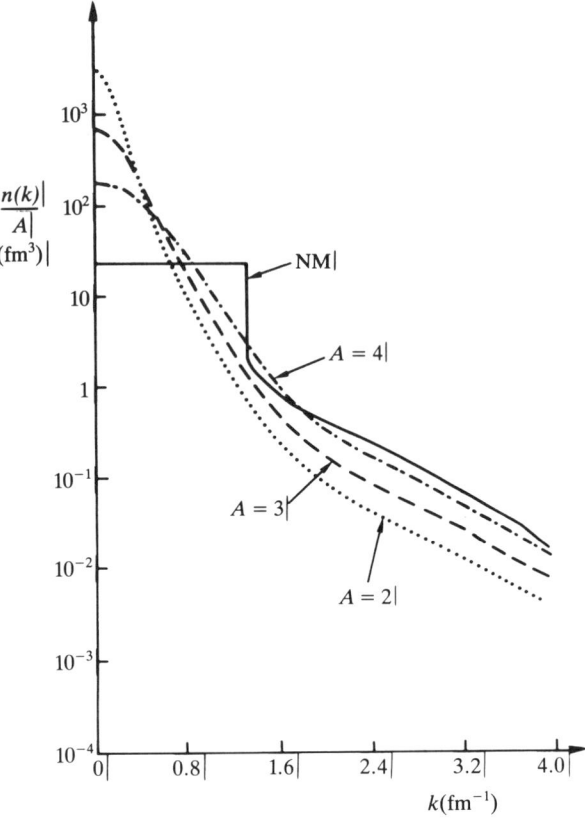

FIG. 6.8. $n(k)/A$ calculated by Schiavilla et al. (1986) in $A = 2$, 3, and 4 nuclei and nuclear matter (NM).

matter. The results presented in Fig. 6.8 for the momentum distributions in ^{3}He and ^{4}He are in qualitative agreement with these from earlier studies (Dieperink et al. 1976; Zabolitzky and Ey 1978). As can be seen, the results for $A = 2$, 3, 4, and nuclear matter are proportional to each other. It is found that $n(k > 2.5 \text{ fm}^{-1})/A$ increases for $A = 2$, 3, 4, and ∞. The observed proportionality of the momentum distributions is in agreement with the result obtained by a phenomenological method based on harmonic-oscillator wavefunctions with short-range and tensor pair correlations (Traini and Orlandini 1985), in which, however, $n(k > 2.5 \text{ fm}^{-1})/A$ decreases as A increases from 4 to 40.

Another variational many-body approach using eqn (6.28) and the correlation operator (6.29) has been applied to the calculation of the momentum distributions of ^{12}C, ^{16}O, and ^{40}Ca by Benhar et al. (1986).

The results have been obtained in the framework of the cluster-expansion technique (Clark 1981) and using the V6 version of the Reid soft-core interaction (Benhar et al. 1979). The momentum distributions of ^{16}O and ^{40}Ca obtained by Benhar et al. (1986) are compared in Figs. 6.9 and 6.10 with the results of the model including short-range and tensor correlations (Traini and Orlandini 1985) (see the next section of this chapter), with the CDFM (Antonov et al. 1979, 1980, 1986b) (see Section 8.3), as well as with the predictions of different Hartree–Fock-type approaches (Elton and Swift 1967; Negele 1970). The result of Benhar et al. (1986) at large k are close to those obtained for ^{16}O in the exp(S) method

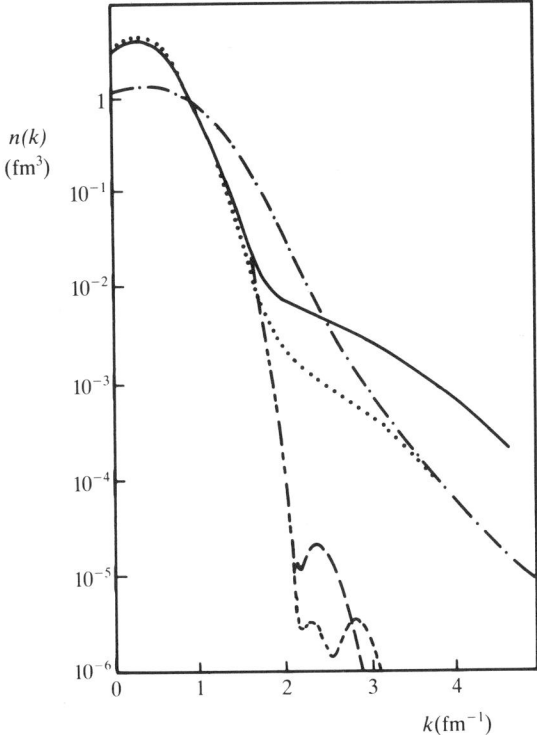

FIG. 6.9. Nucleon momentum distribution in ^{16}O. Solid line: approach of Benhar et al. (1986); long-dashed line: independent-particle model with Elton-Swift (1967) wavefunctions; short-dashed line: with density-dependent Hartree–Fock wavefunctions (Negele 1970); dotted line: results of the model of Traini and Orlandini (1985); dot-dashed line: result of the (CDFM) (Antonov et al. 1980, 1983b, 1986b).

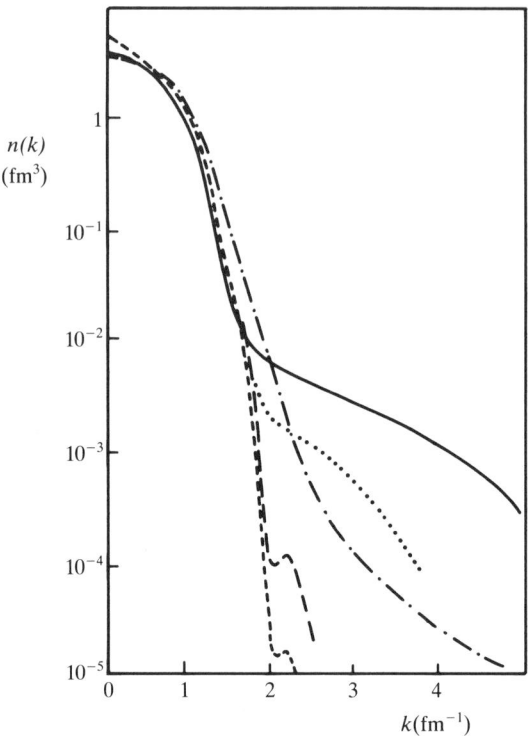

FIG. 6.10. The same as in Fig. 6.9 for ^{40}Ca.

(Zabolitzky and Ey 1978) and in the Brueckner theory of finite nuclei (Van Orden *et al.* 1980).

6.3. Short-range and tensor correlation effects in the two-body density matrix

The inclusion of short-range and tensor correlation effects is rather a complicated problem especially for the microscopic theory of nuclear structure. Several methods have been proposed to treat complex tensor forces and to describe their effects on the nuclear ground state (Thouless 1972; Kuo and Brown 1966; Kuo *et al.* 1971; Moszkowski and Scott 1960; Bethe *et al.* 1963; Villars 1963; Da Providencia and Shakin 1965; Brink and Grypeos 1967).

A simple phenomenological method for introducing dynamical short-range and tensor correlations has been introduced by Dellagiacoma *et al.*

(1983). In this method a two-body correlation operator $U(1,2)$ is introduced to act on the pair wavefunction. Such an approach resembles the earlier approaches construcing the 'exact' wavefunction by means of an operator $\hat{F}$ ($\psi = \hat{F}\Phi$) (Brueckner et al. 1955) or by a correlation Jastrow (1955) and Jastrow-type (Fantoni and Pandharipande 1984) factor $\Pi \hat{f}$ ($\psi = \Pi \hat{f}\Phi$) acting on the uncorrelated determinant wavefunction Φ. In this approach the correlation operator is applied directly in the two-body density matrix written in terms of a Slater-determinant wavefunction.

A similar correlating operator $U(1,2)$, a unitary one, was proposed earlier (Da Providencia and Shakin 1964; Małecki and Picchi 1970, 1971, 1973) for describing the short-range correlation effects.

The two-body density matrix in the case of a Slater determinant Φ is defined as:

$$\rho(v_1, v_2; v_1', v_2') = \sum_{a,b} [\langle v_1, v_2 | ab \rangle \langle ab | v_1' v_2' \rangle - \langle v_1 v_2 | ab \rangle \langle ab | v_2' v_1' \rangle], \quad (6.36)$$

where $v_i \equiv (r_i, s_i^z, t_i^z)$ labels the spatial coordinate, spin, and isospin projection of the ith nucleon.

The correlated two-body density is written as:

$$\rho(v_1, v_2; v_1', v_2') = \sum_{a,b} [\langle v_1 v_2 | u(1,2) | ab \rangle \langle ab | u^+(1,2) | v_1' v_2' \rangle - \langle v_1 v_2 | u(1,2) | ab \rangle \langle ab | u^+(1,2) | v_2' v_1' \rangle]. \quad (6.37)$$

If single-particle wavefunctions of the harmonic-oscillator type are used, the two-particle state $|a(1), b(2)\rangle$ can be expanded on the basis of the relative and centre-of-mass coordinates $r = r_1 - r_2$, $R = \frac{1}{2}(r_1 + r_2)$, the total angular momentum, and spin and isospin of the pair:

$$|a(1)b(2)\rangle = \sum C_{ab} |nlm\rangle |NLM\rangle |SS^z\rangle |TT^z\rangle, \quad (6.38)$$

where $|a(1)b(2)\rangle$ stands for all harmonic-oscillator single-particle quantum numbers, N, L, M; and n, l, m represent harmonic-oscillator radial and angular quantum numbers of the centre-of-mass and relative motions of the pair. S, T are the spin and isospin of the pair, S^z and T^z being their third components. The quantities C_{ab} are expressed by means of the coefficient of the Moshinsky transformation.

The short-range effects can be included in the Slater determinant by defining the operator $u(1,2)_{s.r.}$ that acts on the radial part of the pair wavefunction

$$[u(1,2)]_{\text{s.r.}} |nlm\rangle = \frac{f(r)}{\sqrt{N_{nl}}} |nlm\rangle, \qquad (6.39)$$

$$f(r) \xrightarrow[r \to 0]{} 0,$$
$$f(r) \xrightarrow[r \to \infty]{} 1, \qquad (6.40)$$

and the tensor effect by definition of the $u(1,2)_{\text{tens}}$ operator acting both on the angular and the radial parts of the relative motion of the pair:

$$[u(1,2)]_{\text{tens}} |nlSJJ^z\rangle = \frac{1}{\sqrt{M_{lJ}}} (\delta_{S0} + S_{12}) |nlSJJ^z\rangle, \qquad (6.41)$$

where

$$\boldsymbol{J} = \boldsymbol{L} + \boldsymbol{S}, \qquad (6.42)$$

$$S_{12} = 3(\boldsymbol{\sigma}_1 \cdot \hat{\boldsymbol{r}})(\boldsymbol{\sigma}_2 \cdot \hat{\boldsymbol{r}}) - \boldsymbol{\sigma}_1 \cdot \boldsymbol{\sigma}_2, \qquad (6.43)$$

and N_{nl} and M_{lJ} ensure normalization. For practical applications $f(r)$ is chosen in the form:

$$f(r) = \begin{cases} 1 - a \exp\{-\gamma(\alpha^2/4)(r - r_c)^2\}, & r \geq r_c, \\ 0, & r < r_c. \end{cases} \qquad (6.44)$$

The action of the tensor operator is restricted to deuteron-like states only $(l = 0, S = 1, T = 0)$:

$$[u(1,2)]_{\text{tens}} |n, {}^3S_1, J^z, T = 0\rangle = \sqrt{(1 - \eta^2)} \varphi_{n0}(r) |n, {}^3S_1, J^z, T = 0\rangle$$
$$+ \eta \varphi_{n2}(r) |n, {}^3D_1, J^z, T = 0\rangle. \qquad (6.45)$$

$\varphi_{nl}(r)$ are the radial wavefunctions chosen, for the sake of simplification, to be harmonic-oscillator functions with an oscillator parameter α_D.

Taking into account the relation (1.31) between the two-body and the one-body density matrices the following explicit expressions for the nucleon momentum distribution $n(k)$ and form factor $F(q)$ in the cases of ^{4}He, ^{16}O, and ^{40}Ca are obtained by Traini and Orlandini (1985):

$$^4\text{He:} \quad n(k) = \frac{1}{\alpha^3 \pi^{\frac{3}{2}}} \left\{ \frac{2 - \eta^2}{2N_{10}} \left[\exp\left(-\frac{k^2}{\alpha^2}\right) - \frac{16}{(3\gamma + 4)^{\frac{3}{2}}} \exp\left(-\frac{2k^2}{\alpha^2} \frac{(2 + \gamma)}{(3\gamma + 4)}\right) \right. \right.$$
$$\left. + \frac{1}{(1 + \gamma)^{\frac{3}{2}}(1 + \frac{1}{2}\gamma)^{\frac{3}{2}}} \exp\left(-\frac{2k^2}{\alpha^2} \frac{1}{(2 + \gamma)}\right) \right]$$
$$\left. + \eta^2 \frac{1}{60} \frac{\alpha^3}{\alpha_D^3} \left(\frac{87}{2} - 14 \frac{k^2}{\alpha_D^2} + \frac{2k^4}{\alpha_D^4} \right) \exp\left(-\frac{k^2}{\alpha_D^2}\right) \right\},$$

$$(6.46)$$

PHENOMENOLOGICAL CORRELATION METHODS

^{16}O: $n(k) = \dfrac{1}{16\alpha^3\pi^{\frac{3}{2}}} \Bigg\{ \dfrac{14}{N_{10}} (2-\eta^2)\bigg[\exp\Big(-\dfrac{k^2}{\alpha^2}\Big) - \dfrac{16}{(3\gamma+4)^{\frac{3}{2}}}$

$\times \exp\Big(-\dfrac{2k^2}{\alpha^2}\dfrac{(2+\gamma)}{(3\gamma+4)}\Big) + \dfrac{1}{(1+\gamma)^{\frac{3}{2}}(1+\frac{1}{2}\gamma)^{\frac{3}{2}}}$

$\times \exp\Big(-\dfrac{2k^2}{\alpha^2}\dfrac{1}{(2+\gamma)}\Big)\bigg]$

$+ \eta^2 \dfrac{14}{30} \dfrac{\alpha^3}{\alpha_D^3}\Big(\dfrac{87}{2} - 14\dfrac{k^2}{\alpha_D^2} + 2\dfrac{k^4}{\alpha_D^4}\Big)\exp\Big(-\dfrac{k^2}{\alpha_D^2}\Big)$

$- 8\Big(3 - \dfrac{k^2}{\alpha^2}\Big)\exp\Big(-\dfrac{k^2}{\alpha^2}\Big)\Bigg\}, \hfill (6.47)$

^{40}Ca: $n(k) = \dfrac{1}{40\pi^{\frac{3}{2}}\alpha^3}\Bigg\{\dfrac{185}{4N_{10}}(2-\eta^2)\bigg[\exp\Big(-\dfrac{k^2}{\alpha^2}\Big)$

$- \dfrac{16}{(3\gamma+4)^{\frac{3}{2}}}\exp\Big(-\dfrac{2k^2}{\alpha^2}\dfrac{(2+\gamma)}{(3\gamma+4)}\Big) + \dfrac{1}{(1+\gamma)^{\frac{3}{2}}(1+\frac{1}{2}\gamma)^{\frac{3}{2}}}$

$\times \exp\Big(-\dfrac{2k^2}{\alpha^2}\dfrac{1}{(2+\gamma)}\Big)\bigg] + \eta^2\dfrac{37}{24}\dfrac{\alpha^3}{\alpha_D^3}\Big(\dfrac{87}{2} - 14\dfrac{k^2}{\alpha_D^2} + 2\dfrac{k^4}{\alpha_D^4}\Big)$

$\times \exp\Big(-\dfrac{k^2}{\alpha_D^2}\Big) - \dfrac{1}{2}\Big(165 - 16\dfrac{k^4}{\alpha^4}\Big)\exp\Big(-\dfrac{k^2}{\alpha^2}\Big)\Bigg\},$

$\hfill (6.48)$

where

$$N_{10} = 1 - \dfrac{2}{(1+\frac{1}{2}\gamma)^{\frac{3}{2}}} + \dfrac{1}{(1+\gamma)^{\frac{3}{2}}}. \hfill (6.49)$$

The normalization of $n(k)$ is

$$\int n(\mathbf{k})\,\mathrm{d}\mathbf{k} = 1; \hfill (6.50)$$

^{4}He: $F(q) = \frac{1}{2}[(2-\eta^2)F_B(q^2) + \eta^2 F_A(q^2)], \hfill (6.51)$

^{16}O: $F(q) = \frac{7}{8}[(2-\eta^2)F_B(q^2) + \eta^2 F_A(q^2)]$

$\quad - \Big(\dfrac{3}{4} + \dfrac{1}{8}\dfrac{q^2}{\alpha^2}\Big)\exp\Big(-\dfrac{1}{4}\dfrac{q^2}{\alpha^2}\Big), \hfill (6.52)$

^{40}Ca: $F(q) = \frac{37}{32}[(2-\eta^2)F_B(q^2) + \eta^2 F_A(q^2)]$

$\quad - \Big(\dfrac{21}{16} - \dfrac{1}{4}\dfrac{q^2}{\alpha^2} - \dfrac{1}{80}\dfrac{q^4}{\alpha^4}\Big)\exp\Big(-\dfrac{1}{4}\dfrac{q^2}{\alpha^2}\Big), \hfill (6.53)$

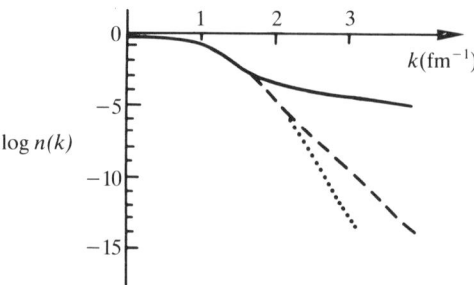

FIG. 6.11. Nucleon momentum distribution of ^{40}Ca (Traini and Orlandini 1985). Dotted line: harmonic-oscillator calculations; dashed line: with only tensor correlations included; solid line: with both tensor and short-range correlations.

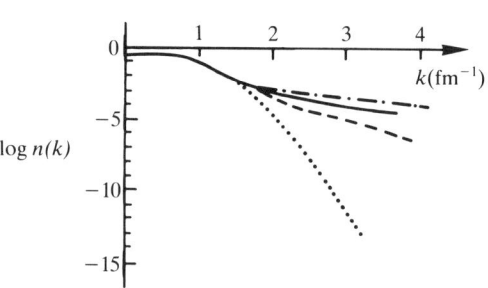

FIG. 6.12. Nucleon momentum distribution of ^{16}O (Traini and Orlandini 1985). Dotted, dashed, and solid lines as in Fig. 6.11. The dot-dashed line is the distribution of Zabolitzky and Ey (1978) corresponding to the RSC potential model.

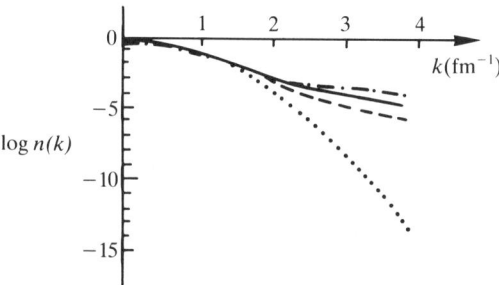

FIG. 6.13. Nucleon momentum distribution of ^{4}He (Traini and Orlandini 1985). Dotted, dashed, and solid lines as in Fig. 6.11. The dot-dashed line as in Fig. 6.12.

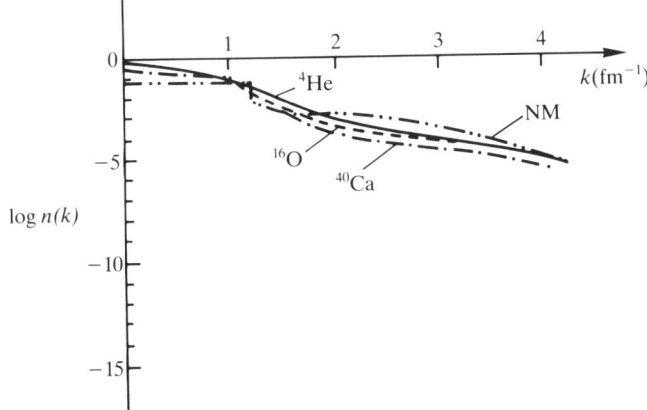

FIG. 6.14. Comparison of momentum distributions of ^{4}He, ^{16}O, and ^{40}Ca evaluated by Traini and Orlandini (1985) and of nuclear matter (Fantoni and Pandharipande 1984).

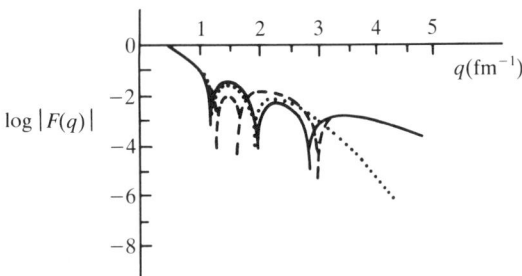

FIG. 6.15. Form factor of ^{40}Ca (Traini and Orlandini 1985). Dotted, dashed, and solid lines as in Fig. 6.11.

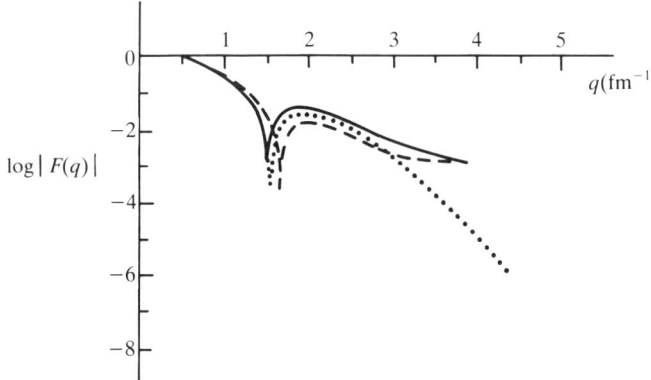

FIG. 6.16. Form factor of ^{16}O (Traini and Orlandini 1985). Dotted, dashed, and solid lines as in Fig. 6.11.

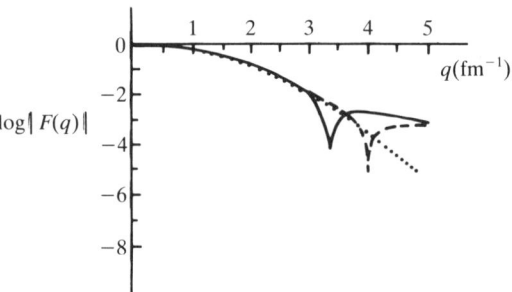

FIG. 6.17. Form factor of ^{4}He (Traini and Orlandini 1985). Dotted, dashed, and solid lines as in Fig. 6.11.

where

$$F_B(q^2) = \frac{1}{N_{10}}\left[B_0(1) - \frac{2}{(1+2\beta)^{\frac{3}{2}}} B_0(y) + \frac{1}{(1+4\beta)^{\frac{3}{2}}} B_0(z)\right], \quad (6.54)$$

$$F_A(q^2) = \frac{1}{15N_{12}}\left\{A_4(1) + 5A_2(1) + \tfrac{15}{4}A_0(1)\right.$$
$$- \frac{2}{(1+\beta)^2(1+2\beta)^{\frac{3}{2}}}\left[\frac{1}{(1+2\beta)^2} A_4(y) + \frac{5}{(1+2\beta)} A_2(y) + \tfrac{15}{4}A_0(y)\right]$$
$$+ \frac{1}{(1+2\beta)^2(1+4\beta)^{\frac{3}{2}}}\left[\frac{1}{(1+4\beta)^2} A_4(z)\right.$$
$$\left.\left. + \frac{5}{(1+4\beta)} A_2(z) + \tfrac{15}{4}A_0(z)\right]\right\}, \quad (6.55)$$

with the notation:

$$\beta = \tfrac{1}{4}\gamma; \quad y = \frac{1+\tfrac{1}{2}\gamma}{1+\tfrac{1}{4}\gamma}, \quad z = \frac{1+\gamma}{1+\tfrac{1}{2}\gamma}, \quad (6.56)$$

$$B_0(x) = \exp\left(-\frac{1}{4x}\frac{q^2}{\alpha^2}\right), \quad (6.57)$$

$$A_0(x) = \exp\left(-\frac{1}{4x}\frac{q^2}{\alpha_D^2}\right), \quad (6.58)$$

$$A_2(x) = \frac{1}{4}\left(6 - \frac{1}{x}\frac{q^2}{\alpha_D^2}\right)\exp\left(-\frac{1}{4x}\frac{q^2}{\alpha_D^2}\right), \quad (6.59)$$

$$A_4(x) = \frac{1}{16}\left(60 - \frac{20}{x}\frac{q^2}{\alpha_D^2} + \frac{1}{x^2}\frac{q^4}{\alpha_D^4}\right)\exp\left(-\frac{1}{4x}\frac{q^2}{\alpha_D^2}\right), \quad (6.60)$$

$$N_{12} = 1 - \frac{2}{(1+\frac{1}{2}\gamma)^{\frac{7}{2}}} + \frac{1}{(1+\gamma)^{\frac{7}{2}}}. \tag{6.61}$$

In eqns (6.51)–(6.53) $F(0) = 1$.

The nucleon momentum distributions calculated by Traini and Orlandini (1985) are shown in Figs. 6.11–6.14 for ^{4}He, ^{16}O, and ^{40}Ca. They are compared with calculations using the $\exp(S)$ method (Zabolitzky and Ey 1978) and with the $n(k)$ of nuclear matter calculated in the variational Jastrow-type method (Fantoni and Pandharipande 1984). As can be seen the effects of tensor correlations are more clearly shown for the light nuclei ^{4}He and ^{16}O than for ^{40}Ca.

The analysis of the presently available results for the high-momentum behaviour of $n(k)$ confirms (see Fig. 6.14) the conclusion drawn in the work of Zabolitzky and Ey (1978), as well as by Ciofi Degli Atti *et al.* (1984), that it is almost the same for $A = 2, 3, 4, 16, 40, \ldots \infty$.

In Figs. 6.15–6.17 the form factors for ^{4}He, ^{16}O, and ^{40}Ca are illustrated. It is seen that the inclusion of the tensor correlations makes an additional minimum appear in ^{4}He and ^{40}Ca.

7

NATURAL-ORBITAL CALCULATIONS OF NUCLEAR DENSITY AND NUCLEON MOMENTUM DISTRIBUTIONS

7.1. Introduction

The simultaneous study of density and momentum distributions of nucleons proves to be a sensitive test of nuclear models. As shown in Chapter 4 one can hardly hope to describe accurately and simultaneously these two quantities in the framework of the independent-particle approach (shell-model, Hartree–Fock calculations etc.). Many plausible features of the single-particle description, however, can be preserved in nuclear approaches using the notion of the natural-orbital representation. These orbitals can be obtained by diagonalizing the one-body density matrix $\rho(r, r')$ resulting from more sophisticated many-particle theories (Brueckner–Hartree–Fock, exp(S), Jastrow methods etc.). In this many of the important short-range and tensor forces are already included in the resulting single-particle (natural) orbitals, as well as in the occupation numbers.

This chapter summarizes the methods that have been used to calculate nuclear density and momentum distributions using the single-particle potential model. This gives the wavefunctions $\psi^p_{JL}(r)$ of the nucleons in the various single-particle states, and the nuclear charge distribution is

$$\rho_{\text{ch}}(r) = \sum_p n_p \, |\psi^p_{JL}(r)|^2, \tag{7.1}$$

where the n_p are the occupation numbers and the sum runs over all protons. In a similar way the nuclear-matter distribution is obtained by summing over the neutrons as well as the protons.

The single-particle wavefunctions can be calculated from a postulated one-body potential or by a self-consistent Hartree–Fock procedure. The former method uses the result that the interaction between a nucleon in the nucleus and all the other nucleons can be represented by a one-body potential with central and spin-orbit terms. The eigenvalues of this potential are the single-particle energies and its eigenfunctions are the wavefunctions of the individual nucleons. If the eigenstates are filled with neutrons and protons starting from the lowest state and respecting the Pauli exclusion principle then, with appropriate parameters, this potential reproduces the nuclear shell structure with the observed magic numbers. In addition, the single-particle energies are in accord with those found experimentally by one-nucleon transfer reactions. The single-particle model thus reproduces many important features of nuclear structure.

This is the essence of the single-particle potential method of calculating nuclear charge and matter distributions. However a rigorous calculation requires the detailed consideration of many effects that can modify the result. Thus for example the potential is usually allowed to be state dependent in order to reproduce the measured single-particle energies; but then the resulting wavefunctions are non-orthogonal and they must be corrected before the density distribution is calculated. The method is therefore described formally in the following section, and used to calculate several nuclear density distributions.

Once the single-particle wavefunctions are known, the corresponding momentum distributions can be obtained by Fourier inversion. Some results of such calculations are given in Section 7.3, and compared with distributions obtained by the Hartree–Fock and related theories, with and without tensor and short-range correlations.

A different way for constructing natural orbitals and occupation numbers is given in Section 7.4. The natural orbitals are obtained from a Saxon-Woods potential with a particular depth-energy dependence. The occupation numbers are specified using RPA and nuclear-matter calculations in which short-range and tensor correlations are taken into account.

7.2. The single-particle potential method: nuclear density distributions

The one-body density matrix and natural orbitals

We briefly review here the formalism needed to express the nuclear density distributions in terms of the one-body density matrix. This will provide the general framework of the single-particle potential (SPP) method.

The total wavefunction $|\psi\rangle$ of a system of A indistinguishable fermions can be expressed in terms of the complete set of Slater determinants, containing all possible particle-hole excitations, constructed from an arbitrary complete set of orthonormal single-particle (s.p.) wavefunctions $\varphi_\alpha(\mathbf{r})$.

The one-body density matrix for this system is expressed by (1.120):

$$\rho(\mathbf{r}, \mathbf{r}') = \sum_{\alpha,\beta} n^1_{\alpha\beta} \varphi^*_\alpha(\mathbf{r}) \varphi_\beta(\mathbf{r}'), \qquad (7.2)$$

where

$$n^1_{\alpha\beta} = \langle \psi | a^+_\alpha a_\beta | \psi \rangle = n^{1*}_{\beta\alpha}. \quad (7.3)$$

Further in this section we shall call the Hermitian matrix $n^1_{\alpha\beta}$ the one-body density matrix. Its diagonal elements are the s.p. occupation numbers

$$n^1_{\alpha\alpha} = \langle \psi | a^+_\alpha a_\alpha | \psi \rangle = \langle n_\alpha \rangle. \quad (7.4)$$

The $\langle n_\alpha \rangle$ range from zero to unity and satisfy the sum rule

$$\text{Tr}(n^1) = \sum_\alpha \langle n_\alpha \rangle = A. \quad (7.5)$$

For this system, any one-body operator $\hat{Q} = \sum_{i=1}^{A} \hat{q}_i$ can be written as

$$\hat{Q} = \sum_{\alpha\beta} q_{\alpha\beta} a^+_\alpha a_\beta, \quad (7.6)$$

with

$$q_{\alpha\beta} = \langle \varphi_\alpha | \hat{q} | \varphi_\beta \rangle. \quad (7.7)$$

The expectation value of $\hat{Q}$ on the exact wavefunction is

$$\langle Q \rangle = \langle \psi | \hat{Q} | \psi \rangle = \sum_{\alpha\beta} q_{\alpha\beta} \langle \psi | a^+_\alpha a_\beta | \psi \rangle = \sum_{\alpha\beta} q_{\alpha\beta} n^1_{\beta\alpha} = \text{Tr}(qn^1). \quad (7.8)$$

Knowledge of n^1 thus enables the expectation value of any one-body operator to be calculated, without explicit knowledge of the total wavefunction $|\psi\rangle$.

Therefore, the one-body density matrix contains all the physical information on the one-body properties of fermion systems.

The n^1 matrix can be diagonalized by a unitary transformation U that defines a new set of single-particle wavefunctions

$$\tilde{\varphi}_\alpha(r) = \sum_\beta U_{\alpha\beta} \varphi_\beta(r). \quad (7.9)$$

The transformed n^1 matrix is

$$\tilde{n}^1_{\alpha\beta} \equiv (U^+ n^1 U)_{\alpha\beta} = \langle \tilde{n}_\alpha \rangle \delta_{\alpha\beta}. \quad (7.10)$$

The s.p. wavefunctions $\tilde{\varphi}_\alpha$ obtained from the transformation that diagonalizes n^1 are called 'natural orbitals'; their importance in the description of many-fermion systems was stressed by Löwdin (1955).

The eigenvalues of the n^1 matrix are the occupation numbers of the natural orbitals; hereafter we will call these 'natural occupation numbers'.

In the natural-orbital basis the expectation value (7.8) of any one-body operator has the simple diagonal form

$$\langle Q \rangle = \sum \tilde{q}_{\alpha\alpha} \langle \tilde{n}_\alpha \rangle, \qquad (7.11)$$

that involves only the occupation numbers and expectation values of $\hat{q}$ in the natural-orbital basis.

The density distribution of a system of A pointlike particles can be easily expressed by using this formalism. The relevant operator is now

$$\hat{\rho}(r) = \sum_{i=1}^{A} \delta(r - r_i) \qquad (7.12)$$

with s.p. matrix elements

$$\rho_{\alpha\beta}(r) = \langle \varphi_\alpha(r') | \delta(r - r') | \varphi_\beta(r') \rangle = \varphi_\alpha^*(r) \varphi_\beta(r). \qquad (7.13)$$

Therefore, the density distribution from eqns (7.8) and (7.11) becomes

$$\rho(r) \equiv \langle \psi | \hat{\rho}(r) | \psi \rangle = \sum_{\alpha\beta} n_{\beta\alpha}^1 \varphi_\alpha^*(r) \varphi_\beta(r) \qquad (7.14)$$

in an arbitrary s.p. basis, and

$$\rho(r) \equiv \langle \psi | \hat{\rho}(r) | \psi \rangle = \sum_\alpha \langle \tilde{n}_\alpha \rangle |\tilde{\varphi}_\alpha(r)|^2 \qquad (7.15)$$

in the natural-orbital basis.

We see that the diagonal expression (7.15) can be used only for natural orbitals; for an arbitrary s.p. basis $\rho(r)$ also contains off-diagonal contributions.

A simple example of this is a bound spinless particle shared between two states with wavefunction of the form

$$\psi(r) = c_1 \varphi_1(r) + c_2 \varphi_2(r), \qquad (c_1^2 + c_2^2 = 1), \qquad (7.16)$$

where c_1, c_2, φ_1, and φ_2 can be taken as real. The occupation numbers are $\langle n_\alpha \rangle = c_\alpha^2$, and the density is

$$\rho(r) = \psi^2(r) = \sum_{\alpha=1}^{2} \langle n_\alpha \rangle \varphi_\alpha^2(r) + 2 c_1 c_2 \varphi_1(r) \varphi_2(r). \qquad (7.17)$$

The natural orbitals here are

$$\tilde{\varphi}_1(r) \equiv \psi(r) = c_1 \varphi_1 + c_2 \varphi_2 \qquad (7.18)$$

with $\langle \tilde{n}_1 \rangle = 1$ and

$$\tilde{\varphi}_2(r) = -c_2 \varphi_1 + c_1 \varphi_2, \qquad (7.19)$$

with $\langle \tilde{n}_2 \rangle = 0$ and for these only the equation

$$\rho(r) = \sum_{\alpha=1}^{2} \langle \tilde{n}_\alpha \rangle \tilde{\varphi}_\alpha^2(r) \qquad (7.20)$$

holds.

Eqns (7.14) and (7.15) have a particularly simple form for systems having total angular momentum $J = 0$. In this case, any choice of s.p. wavefunctions having good angular momentum l, j, m and good parity $(-1)^l$ leads to the partial diagonalization of n^1:

$$J = 0 \rightarrow n^1_{\alpha\beta} \sim \delta_{l_\alpha l_\beta} \delta_{j_\alpha j_\beta} \delta_{m_\alpha m_\beta}. \tag{7.21}$$

This follows from definition (7.3) and the angular momentum and parity orthogonality relations, since now the states $a_\alpha |\psi\rangle$ and $a_\beta |\psi\rangle$ have the same angular momentum and parity as φ_α and φ_β.

This shows that the diagonalization leading to the natural orbitals for a $J = 0$ system has to be carried out only within the $\{ljm\}$ subspaces of the total model space $\{\varphi_\alpha\}$. In other words, for a $J = 0$ system, the transformation (7.9) 'mixes' only s.p. wavefunctions having the same l, j, m, and different principal quantum numbers n.

Moreover, such a system has spherical symmetry, and the density (7.14) is left unaltered after averaging over all directions of space. Writing explicitly

$$\varphi_\alpha(r) = \sum_{\mu\sigma} \langle l_\alpha s \mu \sigma | j_\alpha m_\alpha \rangle \frac{1}{r} y_{n_\alpha l_\alpha j_\alpha}(r) Y^\mu_{l_\alpha}(\Omega) X^\sigma_s, \tag{7.22}$$

and using (7.21), we find (Malaguti et al. 1978)

$$\rho(r) \equiv \frac{1}{4\pi} \int d\Omega \rho(r) = \sum_{lj} \rho_{lj}(r), \tag{7.23}$$

where

$$4\pi r^2 \rho_{lj}(r) = \sum_{nn'} (2j+1) n^1_{n'lj,nlj} Y^*_{nlj}(r) Y_{n'lj}(r), \tag{7.24}$$

in an arbitrary s.p. basis, and

$$4\pi r^2 \rho_{lj}(r) = \sum_n (2j+1) \langle \bar{n}_{nlj} \rangle | \tilde{Y}_{nlj}(r) |^2 \tag{7.25}$$

in a natural-orbital basis. In this case spherical symmetry ensures that the n^1 matrix does not depend on m; this index has, therefore, been dropped and the summation over m gives the $2j + 1$ factor.

To conclude, it is interesting to examine some conditions sufficient to ensure that a particular s.p. basis coincides with a natural-orbital basis. These conditions justify the use of the diagonal density (7.15) in some nuclear models where no explicit reference is made to natural orbitals.

From definition (7.3) of the n^1 matrix it follows:

i) If the total wavefunction of the system is a unique Slater determinant, the s.p. wavefunctions are natural orbitals. This is the case because for $\alpha \neq \beta$, $a_\alpha | \psi \rangle$ and $a_\beta | \psi \rangle$ are different Slater determinants and so their overlap $n^1_{\alpha\beta} = 0$. This is the case of the Hartree-Fock model.

ii) If, for a $J = 0$ system, all s.p. states except at most one within each $\{ljm\}$ subspace are either completely full or completely empty, then the s.p. wavefunctions are natural orbitals.

This condition, already discussed by Malaguti et al. (1978), can be shown as follows: the matrix element $n^1_{\alpha\beta} = \langle\psi|a^+_\beta a_\alpha|\psi\rangle \neq 0$ for $\alpha \neq \beta$ means that a particle can be shifted from state α to β, and this means $\langle n_\alpha \rangle \neq 0$, $\langle n_\beta \rangle \neq 1$. But, if $n^1_{\alpha\beta} \neq 0$, then $n^{1*}_{\alpha\beta} = \langle\psi|a^+_\alpha a_\beta|\psi\rangle \neq 0$, and this means $\langle n_\beta \rangle \neq 0$ and $\langle n_\alpha \rangle \neq 1$. Thus both $\langle n_\alpha \rangle$, $\langle n_\beta \rangle \neq 0$ and $\langle n_\alpha \rangle$, $\langle n_\beta \rangle \neq 1$. For a $J = 0$ system, all this can be repeated within each $\{ljm\}$ subspace (7.21). Therefore, if there is only one s.p. state in each subspace whose occupation number is neither 0 nor 1, then $n^1_{\alpha\beta} = 0$ for any $\alpha \neq \beta$.

This is the case of the nuclear shell-model constructed from a 'closed core' and some 'valence particles': $|\psi\rangle$ is no longer approximated by a single Slater determinant and some 'configuration mixing' is introduced. But this mixing is sufficiently weak to involve not more than one s.p. state for each $\{ljm\}$ subspace. An example is provided by the 's-d shell model', where active nucleons occupy the $2s_{\frac{1}{2}}$, $1d_{\frac{5}{2}}$, $1d_{\frac{3}{2}}$ shells outside a closed ^{16}O core.

iii) If all elementary configurations (Slater determinants) entering in the total wavefunction of the system differ by more than the shift of one particle, then the s.p. wavefunctions are natural orbitals. This is because $n^1_{\alpha\beta} = \langle\psi|a^+_\beta a_\alpha|\psi\rangle \neq 0$ for $\alpha \neq \beta$ implies that, within $|\psi\rangle$, at least two configurations differ by a 1-particle, 1-hole excitation.

This is approximately the case for the ground state of $J = 0$ even-even nuclei, where the excitations constructed from the simple 'all nucleons coupled to pairs' configuration are more likely to be 2-particles-2 holes, 4p − 4h, 6p − 6h, ..., than 1p − 1h, 3p − 3h, ...

Application to the calculation of charge densities

i) General features of the SPP method

From the properties of the n^1 matrix, the natural orbitals appear as the best s.p. wavefunctions for the calculations of nuclear density distributions and in general of the expectation value of any one-body operator.

Unfortunately, the practical use of natural orbitals is limited by the fact that a method to calculate them from the fundamental nucleon-nucleon force has not yet been worked out in detail (see for example Schäfer and Weidenmüller 1971). In the SPP method an empirical approach to the problem is tried, observing that the natural orbitals should be similar to the wavefunctions used in the single-particle shell-model, based on a

Saxon–Woods potential with little or no configuration mixing. This assumption is justified by the success of the density distributions calculated by Elton and Swift (1967) and Elton and Webb (1970), and by conditions i) to iii) described above. The ground-state density distributions of $J = 0$ even-even nuclei are calculated using Saxon–Woods wavefunctions and the diagonal expression (7.25).

The model space $\{\varphi_\alpha\}$ is limited to the bound states for which one-nucleon transfer reactions show a significant occupancy. Therefore, the contribution to the density coming from continuum single-particle states, whose occupancy may be non-zero as a result of correlations among particles, is neglected. Some of the parameters of the Saxon–Woods potential are varied to get a best fit to the experimental density distributions. The best Saxon–Woods approximation to natural orbitals, within the assumed model space, is found in this way.

After that, one can go beyond this approximation and explicitly construct natural orbitals by mixing the Saxon–Woods functions contained in the model space according to eqns (7.9)–(7.10).

In more detail the natural radial functions $\bar{y}_\alpha(r)$ in eqn (7.25) are approximated by the radial eigenfunctions $y_\alpha(r)$ of a local Saxon–Woods potential of standard form and depth V_L,

$$V(r) = -V_L f_1(r) + V_S \left(\frac{\hbar}{m_\pi c}\right)^2 \frac{1}{r} \frac{df_2(r)}{dr} \boldsymbol{L} \cdot \boldsymbol{\sigma} + V_C(r), \qquad (7.26)$$

where

$$f_{1,2}(r) = \{1 + \exp[(r - R_{1,2})/a_{1,2}]\}^{-1}, \qquad (7.27)$$

with $R_1 = RA^{\frac{1}{3}}$, $R_2 = 1.1 A^{\frac{1}{3}}$ and $a_2 = 0.65$ fm.

R and a_1 are treated as free parameters in the optimizing calculations. V_C (only for protons) is the Coulomb potential of a uniformly charged sphere having the experimentally measured r.m.s. charge radius, while V_L and V_S are chosen so as to reproduce the s.p. energies ε_α of the $j = l \pm \frac{1}{2}$ doublet, measured in one-nucleon transfer reactions. If only the centroid of the doublet is known, the spin-orbit potential V_S is set to 7 MeV (Millener and Hodgson 1973; Malaguti and Hodgson 1973). According to Clement (1969) and Baranger (1970), the experimental binding energy E_α for the s.p. level α is defined by the two-sided centroid

$$E_\alpha \equiv -\varepsilon_\alpha = \left(\sum_N S_{N\alpha}(E_N - E_0) + \sum_n S_{n\alpha}(E_0 - E_n)\right) \bigg/ (2j_\alpha + 1), \quad (7.28)$$

where E_0 is the ground-state energy of the target nucleus A, E_N and E_n are the energies of states in nuclei $A - 1$ and $A + 1$, respectively, $S_{N\alpha}$ and $S_{n\alpha}$ their s.p. spectroscopic factors, measured by pick-up and stripping reactions from or to the s.p. level α, respectively, and normalized by the sum rule

$$\sum_N S_{N\alpha} + \sum_n S_{n\alpha} = 2j_\alpha + 1. \tag{7.29}$$

The use of simple one-sided centroids is probably more popular among experimentalists, but definition (7.28) gives energies that are better fitted by potential (7.26) for non-closed shells (Malaguti 1978).

The natural occupation numbers $\langle \bar{n}_\alpha \rangle$ to be used in eqn (7.25) are approximated by those extracted from one-nucleon transfer reactions according to the French–MacFarlane (1961) sum rules:

$$\langle \bar{n}_\alpha \rangle \approx \langle n_\alpha \rangle = \left(\sum_N S_{N\alpha} \right) \bigg/ (2j_\alpha + 1) = 1 - \left(\sum_n S_{n\alpha} \right) \bigg/ (2j_\alpha + 1) \tag{7.30}$$

for states near the Fermi-level, and are simply taken to be unity for deep-lying orbits.

The assumption near the Fermi-level is partially justified by the work of Clement (1973) who derives an expression for the density similar to eqn (7.25). If the centre-of-mass correction is neglected, Clement's formula is

$$4\pi r^2 \rho_{lj}(r) = \sum_N S_{Nlj} \Phi^2_{Nlj}(r), \tag{7.31}$$

where $\Phi_N(r)$ is a radial overlap function, i.e. the radial part of a s.p. function describing the overlap between the ground state of nucleus A and the Nth state of nucleus $A-1$, excited in a one-nucleon pick-up reaction. In principle, the overlap functions depend on the index N, but as Clement (1969) points out, for a group of levels with energies E_N corresponding to removing a nucleon from a s.p. level near the Fermi-surface the N dependence is weak and from (7.30)

$$4\pi r^2 \rho_{lj}(r) \approx \Phi^2_{lj}(r) \sum_N S_{Nlj} = \Phi^2_{lj}(2j+1)\langle n_{lj} \rangle. \tag{7.32}$$

Therefore, for a group of levels of this kind, Φ_{lj} can be approximately identified with a natural orbital and $\sum_N S_{Nlj}/(2j+1)$ with the corresponding natural occupation number.

Finally, since the physical one-body potential is expected to be non-local, the use of the local potential (7.26) appears questionable; a constant non-local potential seems to be a better choice. Following common practice, non-locality effects are introduced into wavefunctions $y_{nlj}(r)$ by multiplying them by the Perey factor (Perey 1963a, 1963b):

$$y_{nlj}(r) \rightarrow y_{nlj}(r) \left[1 + \frac{\mu \beta^2}{2\hbar^2} V_L f_1(r) \right]^{-\frac{1}{2}}, \tag{7.33}$$

where μ is the reduced nucleon mass and $\beta \approx 1$ fm the non-locality range that is treated as a free parameter in the fitting calculations.

The SPP method has been applied to the proton states of a series of nuclei, and the resulting charge distributions have been fitted to the model-independent ones obtained from electron scattering. This fitting has been carried out by varying the parameters: R (defining the radius of the central potential as $R_1 = RA^{\frac{1}{3}}$); a_1, the diffuseness of the central potential; β, the non-locality range of the central potential.

In some cases, additional free parameters were provided by one or two occupation numbers or a single-particle energy whose experimental value suffered from large uncertainty. In general, the calculated charge density is sensitive to the occupation numbers, but depends only weakly on the single-particle energies.

For accurate calculations several refinements and corrections have been made to the simple formalism which we shall now discuss.

ii) Folding the finite nucleon-charge distribution

Nucleons possess an internal structure and, in particular, a charge distribution. Therefore, eqn (7.25) must be folded with a function describing their internal charge density. For present purposes, it is sufficiently accurate to use a sum of 3-dimensional gaussians with spherical symmetry

$$h(r) = \sum_k d_k \left(\frac{1}{a_k \sqrt{\pi}}\right)^3 \exp\left(-\frac{r^2}{a_k^2}\right), \qquad (7.34)$$

where the parameters d_k and a_k were found by Chandra and Sauer (1976) from their analysis of electron-nucleon scattering. For protons they are

$$d_{1,2,3} = 0.506, 0.328, 0.166 \text{ fm}; \qquad a_{1,2,3} = 0.657, 0.373, 1.325 \text{ fm} \qquad (7.35)$$

and correspond to a r.m.s. charge radius of 0.88 fm.

Neutrons also, though their total charge is zero, have a charge distribution with a mean-square radius of -0.116 fm^2. It was found by Bertozzi et al. (1972) that the apparently anomalous decrease of the charge radius from ^{40}Ca to ^{48}Ca is partly due to the $f_{\frac{7}{2}}$ neutrons. The neutron charge density is described by eqn (7.34) and (Chandra and Sauer 1976):

$$d_{1,2} = +1, -1; \qquad a_{1,2} = 0.685, 0.739 \text{ fm}. \qquad (7.36)$$

Folding (7.25) with (7.34) and performing angular integrals yields again (7.25) with a modified form factor (Malaguti et al. 1978)

$$\left|\frac{y_\alpha(r)}{r}\right|^2 \to \int_0^\infty dr'\, g(r, r') \left|\frac{y_\alpha(r')}{r'}\right|^2, \qquad (7.37)$$

with

$$g(r, r') = \frac{\pi r'}{r} \sum_k \{\pi^{-\frac{3}{2}} d_k a_k^{-1} \left[\exp\left[-\left(\frac{r-r'}{a_k}\right)^2 \right] - \exp\left[-\left(\frac{r+r'}{a_k}\right)^2 \right] \right]\}. \tag{7.38}$$

iii) Centre-of-mass motion

The density distributions are measured in a reference frame in which the centre of mass (c.m.) of the nucleus is fixed. However, like in all shell-model or Hartree-Fock calculations, it is the centre of potential that is fixed, and this introduces a spurious c.m. vibration around the origin, resulting in the density, $\rho_{f.p.}$, having a larger radial extent than that seen in the c.m. frame, $\rho_{c.m.}$. It is therefore necessary to derive the relation between $\rho_{c.m.}$ and $\rho_{f.p.}$. While, in general, this relation is difficult to derive (Lipkin 1958), in the particular case of the ground state in a harmonic-oscillator potential it has been found by Elliot and Skyrme (1953):

$$\rho_{c.m.}(r) = \left(b\sqrt{\frac{\pi}{A}} \right)^{-3} \int dr' \exp[A \, |r - r'|^2/b^2] \rho_{f.p.}(r'), \tag{7.39}$$

where $b = (\hbar/m\omega)^{\frac{1}{2}}$, m is the nucleon mass, ω the oscillator frequency, and A the nuclear mass number.

After this correction, $\rho_{c.m.}$ has to be folded again with the internal density of the nucleon. It is here that form (7.34) for the nucleon density appears very useful. A straightforward calculation shows that unfolding (7.39) plus a finite-size correction is equivalent to a single finite-size folding, but with a reduced range parameter

$$a_k^2 \rightarrow a_k^2 - b^2/A. \tag{7.40}$$

This correction is exact for a harmonic-oscillator potential, and may be used also for Saxon–Woods proton densities; this approximation gives reasonable results for electron scattering up to $q_{max} \approx 3 \, \text{fm}^{-1}$ (Gamba et al. 1973). For each nucleus the parameter b is determined by the condition that a simple harmonic-oscillator model of the nucleus (without configuration mixing) reproduces the measured r.m.s. charge radius.

iv) Relativistic corrections

The high precision of experimentally measured charge densities makes it necessary to consider also the small ($\approx 1\%$) relativistic corrections. Among these, the oldest and best known are the ones to the electron-nucleus interaction which arise from reducing the nucleon's 4-component s.p. current to a 2-component (non-relativistic) form. These have been described by de Forest and Walecka (1966) and worked out in detail for a $J = 0$ nucleus by Bertozzi et al. (1972) and by Chandra and Sauer (1976). The corrections are of two types, called the 'Darwin–Foldy term' and the 'spin-orbit contribution'. The Darwin–Foldy term can be combined with the Gaussian folding of the proton size and the c.m. motion by using eqn (7.34) with parameters (7.35) corrected as (see also eqn (7.40))

$$a_k^2 \to a_k^2 - b^2/A + 2.21 \times 10^{-2} \text{ fm}^2. \tag{7.41}$$

The spin-orbit correction results in an 'effective' contribution to the charge density that can be written as (Campi et al. 1974):

$$\rho_{\text{s.o.}}(r) = -\tfrac{1}{2}\left(\frac{\hbar}{mc}\right)^2 \frac{1}{r^2}\frac{d}{dr}\left[r \sum_{lj\tau} \langle n_{lj\tau}\rangle \mu_\tau \langle \boldsymbol{\sigma} \cdot \boldsymbol{l}\rangle_{lj} \rho_{lj\tau}\right]. \tag{7.42}$$

v) Wavefunction orthogonalization

The formalism described in this section is correct only if the s.p. wavefunctions form a complete orthonormal set. Since the Saxon–Woods potential is fitted to each s.p. energy in turn, different potentials are actually used and orthogonality is not ensured. For states with different l or j or m, orthogonality is ensured by the angular part of the wavefunctions, but within the same $\{ljm\}$ subspace it is not guaranteed.

Of course, it is possible to modify this formalism so as to work with non-orthogonal wavefunctions, but in practice it is easier to perform an algebraic transformation on the wavefunctions of each $\{ljm\}$ subspace that makes them orthogonal. This change of basis is arbitrary, provided one transforms the occupation numbers accordingly. This is done using the Gram-Schmidt method starting from the most bound level and following the order of decreasing binding energy. The details are given by Malaguti et al. (1978) including the transformation of the occupation numbers. This shows that, under the general conditions met by these calculations, all transformed occupation numbers are unity except possibly that of the least bound state in each subspace. In nearly all cases of practical interest, the $\{ljm\}$ subspaces contain two levels (the first exception in the periodic table is the 1, 2, $3s_{\frac{1}{2}}$ levels in ^{208}Pb), thus the transformation may be written as

$$\langle n_1 \rangle = 1, \quad \langle n_2 \rangle = 1 - (1 - \langle \bar{n}_2 \rangle)/|u_{22}|^2, \tag{7.43}$$

where the bar in $\bar{n}_2$ means 'before orthogonalization.' It was shown by Malaguti et al. (1978) that $\langle n_2 \rangle$ becomes

$$\langle n_2 \rangle = \frac{\langle \bar{n}_2 \rangle - |o_{12}|^2}{1 - |o_{12}|^2} \approx \langle \bar{n}_2 \rangle + |o_{12}|^2(\langle \bar{n}_2 \rangle - 1) \tag{7.44}$$

to second order in o_{12}, the overlap between the two wavefunctions before orthogonalization. This overlap can be large (up to ≈20%) when the wavefunctions are first generated in potential (7.26), but it is then strongly reduced by the non-locality correction (7.33). The reason is that the Perey factor corresponds to a unique non-local potential (hence orthogonal wavefunctions) and its local equivalent depends linearly on energy. Typical values for o_{12} after the Perey correction (7.33) range from 1% to 6% and, since $\langle \bar{n}_2 \rangle - 1$ is of order unity, from (7.44)

$$\langle n_2 \rangle \approx \langle \bar{n}_2 \rangle \tag{7.45}$$

is a very good approximation.

The non-local single-particle wavefunctions are only slightly non-orthogonal with maximum overlap ≈6%. Orthogonalization changes them linearly in the overlap; in principle, the occupation numbers defined by eqn (7.30) should also be changed but the correction is quadratic in the overlap and can be neglected.

vi) Energies of the deep states

The spectroscopic data needed for the application of eqn (7.28) exist only for states near the Fermi-surface. For medium-weight nuclei (up to ^{58}Ni) one can make some use of (p, 2p) knock-out reactions, but thereafter (and for neutron states) experimental energies of the deep levels are unknown.

The local potentials for a given nucleus have an approximately linear dependence on energy, corresponding to an approximately constant non-local potential. Starting from the known s.p. binding energies E, the local potential depths V_L are extracted corresponding to eqns (7.26)–(7.27) with 'reasonable' parameters $R = 1.25$ fm and $a_1 = 0.65$ fm. The corresponding non-local potential depths U are then calculated from the relation (Perey 1963a, 1963b):

$$U = V_L \exp\left[\frac{\mu \beta^2}{2\hbar^2}[V_L - E - V_C(0)]\right], \tag{7.46}$$

where $\beta = 0.85$ fm and $V_C(0) = 3Ze^2/2R_C$ (only for protons) is the Coulomb potential at the nuclear centre. The U's are then averaged and the average non-local potential is used to calculate the unknown energies by a numerical solution of eqn (7.46) for E. (V_L is a function of E and this makes the equation implicit.)

The average non-local potential differs for neutrons and protons by the known asymmetry effect and can be parametrized as

$$U_{p,n} = U_0 \pm \frac{N-Z}{A} U_1, \tag{7.47}$$

where the $+$ $(-)$ refers to protons (neutrons).

108   NATURAL-ORBITAL CALCULATIONS

Subsidiary calculations have shown that the uncertainties introduced by these procedures have a negligible effect on the calculated charge densities.

The single-particle potential method has been used to calculate the charge distributions of several nuclei. The first detailed analysis was made for ^{58}Ni, as this was the heaviest nucleus for which all the single-particle binding energies were known. Subsequently analyses were made of the charge densities of ^{40}Ca, ^{39}K, and ^{48}Ca and the results are compared with the experimental charge densities in Fig. 7.1. It is notable that the significantly different radial distributions in the nuclear interior are all

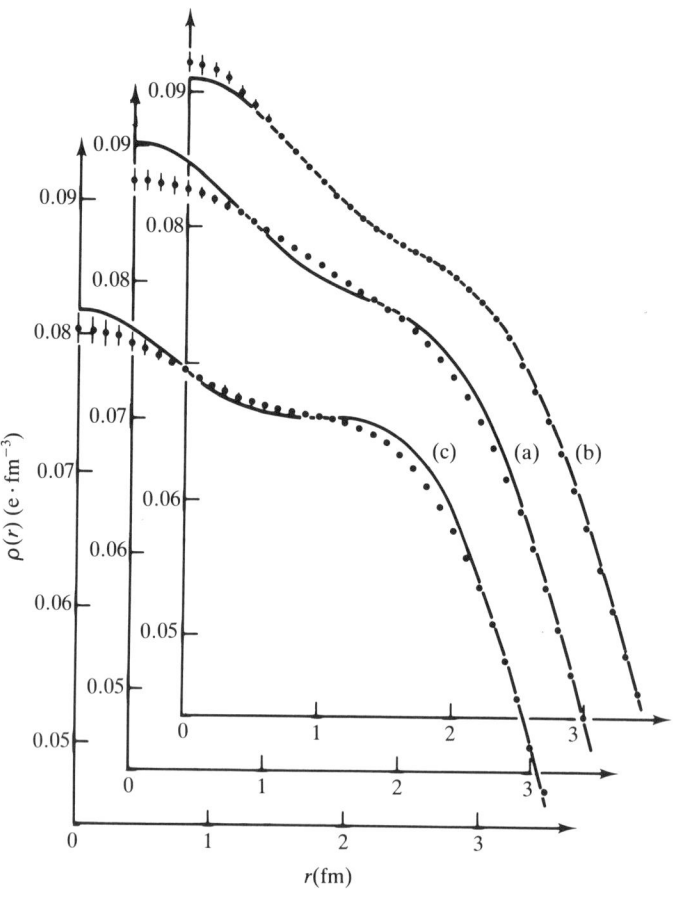

FIG. 7.1. Comparison between the model-independent charge densities of Sick (1974) and the results of best-fit calculations of Malaguti et al. (1982) for a) ^{40}Ca, b) ^{39}K, and c) ^{48}Ca.

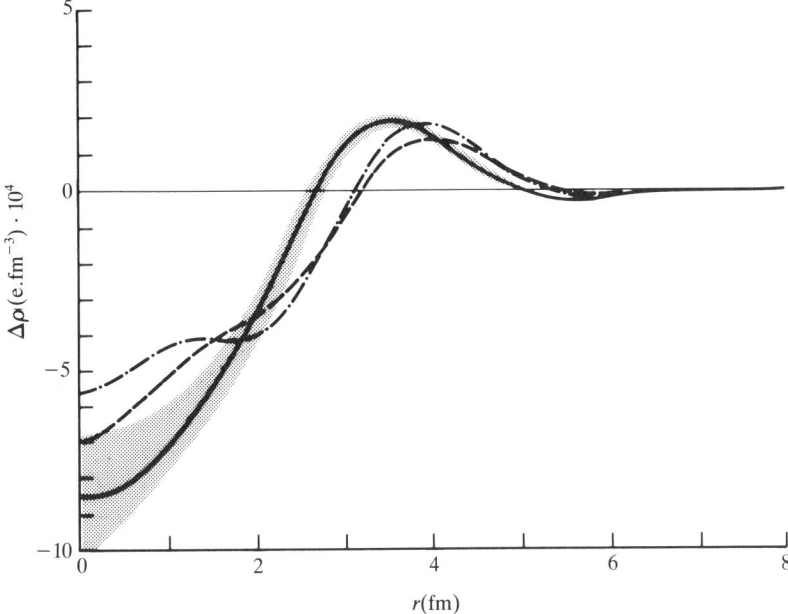

FIG. 7.2. Comparison of experimental charge-density differences $\Delta\rho$ between ^{48}Ca and ^{40}Ca (Sick 1974; Sick *et al.* 1979) and the results of different types of calculation: Experimental results are represented by the shaded area; the Hartree–Fock by the dashed line; Hartree–Fock plus RPA by the dot-dashed line; and Malaguti *et al.* (1982) by the solid line.

quite well given by the theory. For ^{39}K and ^{48}Ca the calculations gave about 0.1 and 0.15 for particles in the $2p_{\frac{3}{2}}$ orbit, respectively.

The charge-density difference between two isotopes such as ^{40}Ca and ^{48}Ca provides a critical test of any theoretical calculation. Take, for example, the self-consistent calculations of Gogny and Grammaticos (see Frois 1979). While they improved fits to individual densities, they were not however able to fit the charge-density difference, even if RPA correlations were included. However as shown in Fig. 7.2 the SPP method is able to fit the charge-density difference between ^{40}Ca and ^{48}Ca, without further parameter variation.

A further test of the model is provided by the charge densities of the zirconium isotopes, and in particular by their charge distribution differences. It was found that the densities can be reproduced by slight variations of the occupation numbers for the p and g orbits, and the results are compared with the experimental densities in Fig. 7.3. The values for the radius and diffuseness of the central potential were found to increase along the isotopic sequence. This may be connected with the

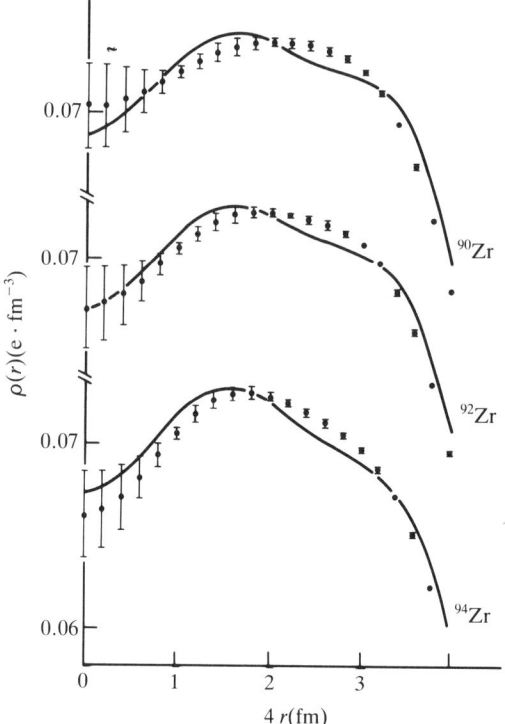

FIG. 7.3. Comparison between the experimental charge density of zirconium isotopes and best-fit calculations of Malaguti *et al.* (1982).

macroscopic effects pointed out by Rothhaas (1978), the change in diffuseness corresponding to the change in the intrinsic deformation and the increase of the potential radius to the macroscopic-core polarization effects following the addition of neutron pairs. This calculation is supported by the results shown in Fig. 7.4 in which the change in R is shown to give extra charge to the surface region and a uniform reduction in the interior. The figure also shows the contribution to the charge-density differences due to these causes and, for ^{92}Zr and ^{94}Zr, to the difference in the p-orbit occupation numbers.

As a final example of the application of the single-particle potential method to the calculations of charge densities we consider ^{208}Pb, which has been the subject of many experimental and theoretical studies. Experimentally, the more recent model-independent analysis of Frois *et al.* (1977) shows less oscillating structure in the nuclear interior than the older work of Friar and Negele (1973, 1975) and of Sprung *et al.* (1976). The distribution of Frois *et al.* (1977) is shown in Fig. 7.5, together with

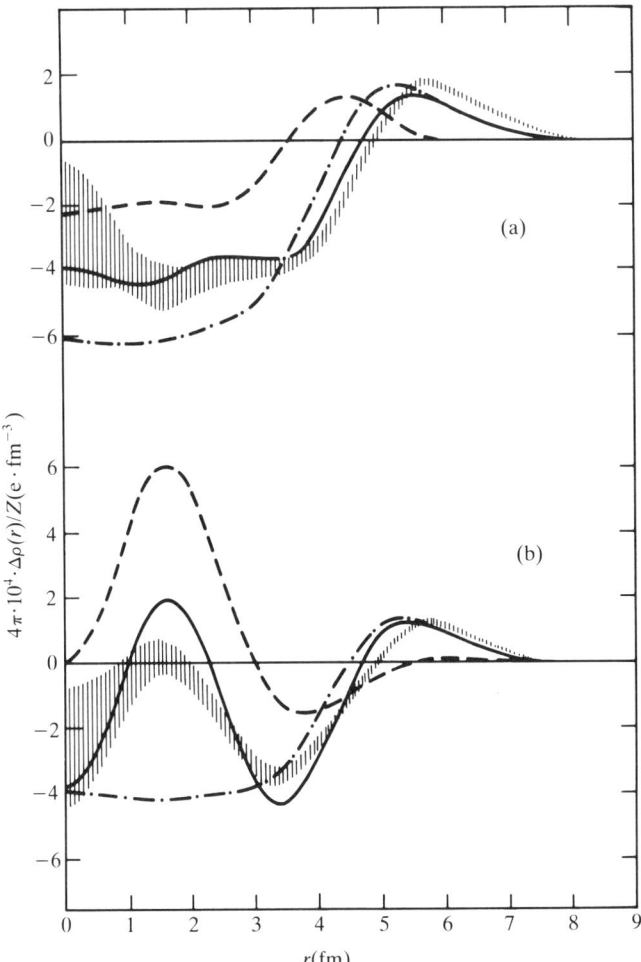

FIG. 7.4. Contributions to charge-density differences in zirconium isotopes (Malaguti *et al.* 1982). The dash-dotted curves show the effect of the radial expansion. The dashed curves show the effect of the difference in best-fit diffuseness for ^{90}Zr–^{92}Zr (a) and the effect of a 0.3 particle transfer from the 1g to the 2p orbit for ^{94}Zr–^{92}Zr (b). The cumulative effects are shown by the solid lines.

the results of Hartree–Fock, RPA, and SPP calculations. It is notable that all these calculations give too much oscillation in the nuclear interior. Subsequent work by Brown *et al.* (1983) showed that a greatly improved fit can be obtained by combining the SPP method for the nuclear surface region with Hartree–Fock theory and Skyrme interactions in the nuclear interior; the results for both the charge and neutron

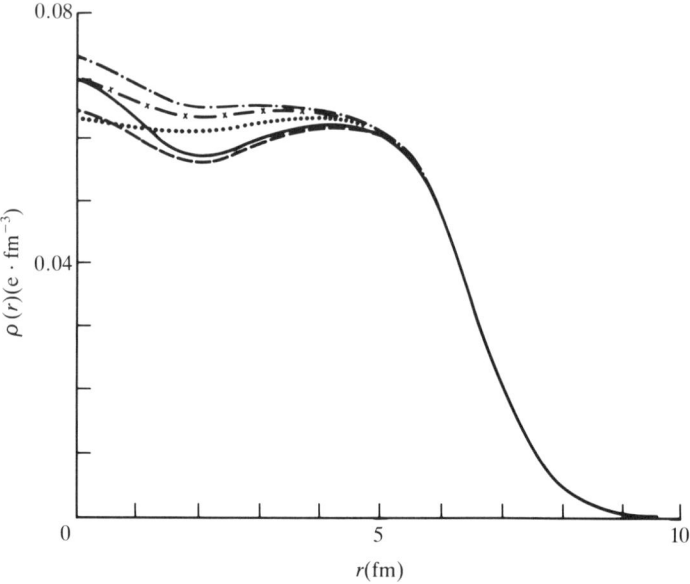

FIG. 7.5. Charge-density distributions in ^{208}Pb. The points show the model-independent density of Frois *et al.* (1977). The RPA(HF) curves correspond to the Hartree–Fock calculations of Frois (1979) with (without) RPA corrections. The SPP curves (Malaguti *et al.* 1982) show the density calculated with a) $n_{3s_{\frac{1}{2}}} = 2$; b) $n_{3s_{\frac{1}{2}}} = 1.75$; $n_{1h_{\frac{9}{2}}} = 0.25$. Experimental data is represented by the dotted line; RPA by the crossed-dashed line; HF by the dot-dashed line; SPP(a) the solid line, and SPP(b) the dashed line.

densities are shown in Fig. 7.6. These calculations make it possible to understand the physical origin of the density oscillations. Starting with a potential with a flat interior, such as a Fermi shape, the oscillations of the proton and neutron densities are almost exactly out of phase because of the different shell structures associated with the magic numbers 82 and 126. Also, because of the strong proton-neutron interaction, the proton potential is determined primarily from the neutron density and vice-versa. Hence, for a zero-range two-body interaction, the Hartree–Fock potential is out of phase with the density, which tends to damp the oscillations in a self-consistent calculation. On the other hand, a sufficiently long-ranged interaction can produce a flat potential by washing out the potential oscillations and a further increase in the range of the potential can produce oscillations in the potential in phase with the density.

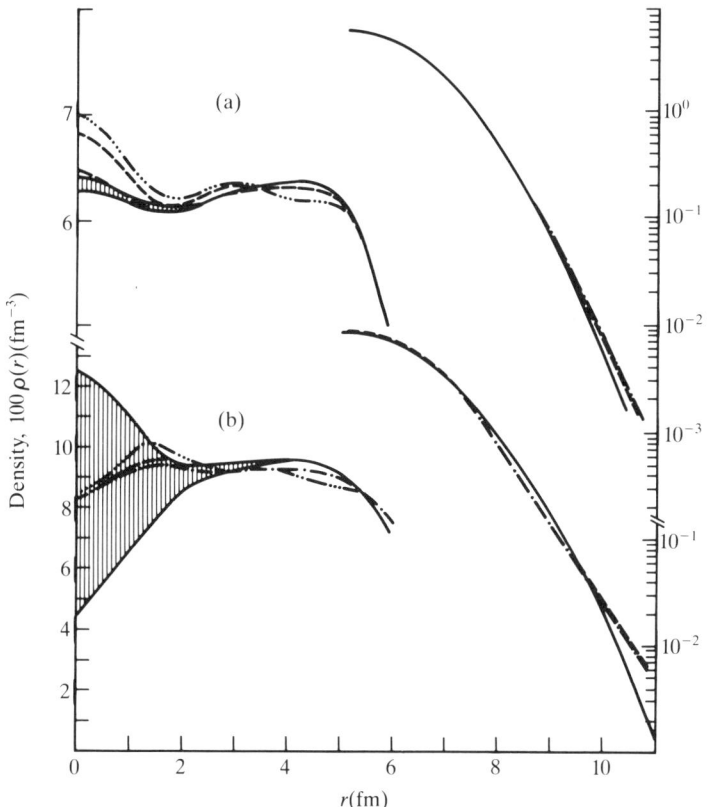

FIG. 7.6. Charge (a) and point neutron (b) densities in ^{208}Pb. The experimental data (shaded) are compared with pure Saxon–Woods calculations (dash three dots) and the mixed Saxon–Woods–Skyrme calculations with Skyrme effective mass M (0.95) (dot dash) and M (0.60) (dash) (B. A. Brown *et al.* 1983).

7.3. The single-particle potential method: nucleon momentum distributions

The single-particle potential method described in the previous section can be used to calculate the nucleon momentum distributions. As shown previously, the nuclear density distribution is given by

$$\rho(r) = \frac{1}{4\pi} \sum_{lj} (2j+1) \sum_n \tilde{n}_{nlj} |\tilde{R}_{nlj}(r)|^2, \tag{7.48}$$

where $\tilde{n}_{nlj}$ are occupation numbers and $\tilde{R}_{nlj}(r)$ the radial wavefunction (see eqn (7.25): $\tilde{R}_{nlj}(r) = \tilde{y}_{nlj}/r$).

The corresponding nucleon momentum distribution is given by

$$n(k) = \frac{1}{4\pi} \sum_{lj} (2j+1) \sum_n \tilde{n}_{nlj} |\tilde{R}_{nlj}(k)|^2, \qquad (7.49)$$

where

$$\tilde{R}_{nlj}(k) = \left(\frac{2}{\pi}\right)^{\frac{1}{2}} (-i)^l \int_0^\infty dr\, r^2 j_l(kr) \tilde{R}_{nlj}(r) \qquad (7.50)$$

and $j_l(kr)$ are spherical Bessel functions of order l. The natural-orbital wavefunctions $\tilde{R}_{nlj}(r)$ are obtained from the Saxon–Woods wavefunctions by the relation (7.9). Since this transforms wavefunctions of differing (l, j, m) independently, the transformation is applied only to states of the same (l, j, m) and differing principal quantum number n. For example in the case of ^{40}Ca it is necessary to transform only the $(1s_{\frac{1}{2}}, 2s_{\frac{1}{2}})$ and $(1p_{\frac{3}{2}}, 2p_{\frac{3}{2}})$ wavefunctions, and in each case this is done by a simple rotation. The occupation numbers, as mentioned in the previous section, are practically unaffected by the transformation.

This method of calculating nucleon momentum distributions has been applied to ^{40}Ca (Antonov et al. 1987) using firstly the occupation numbers of Malaguti et al. (1982) and then some other sets chosen to give approximately the same proportions of protons above the Fermi-sea as were given in the calculations of Jaminon et al. (1986). This enables the effects of proton promotion above the Fermi-sea on the nucleon momentum distribution to be studied. The occupation numbers used in the SPP calculations are given in Table 7.1.

The results of SPP calculations are shown in Fig. 7.7 (curves iii)–v)). In the same figure are given the momentum distributions from the calculations of Traini and Orlandini (1985) including short-range and tensor correlations (curves i), ii), vii)), the result of the coherent fluctuation

TABLE 7.1. Occupation numbers of ^{40}Ca used in SPP calculations.

State n,l,j	(1) $\tilde{n}_{nlj}$ (Malaguti et al. 1982)	(2) $\tilde{n}_{nlj}$
$2p_{\frac{3}{2}}$	0.0375	0.275
$1f_{\frac{7}{2}}$	0.07	0.375
$1d_{\frac{3}{2}}$	0.8975	0.40
$2s_{\frac{1}{2}}$	0.850	0.75
$1d_{\frac{5}{2}}$	1.0	0.80
$1p_{\frac{1}{2}}$	1.0	1.0
$1p_{\frac{3}{2}}$	1.0	1.0
$1s_{\frac{1}{2}}$	1.0	1.0

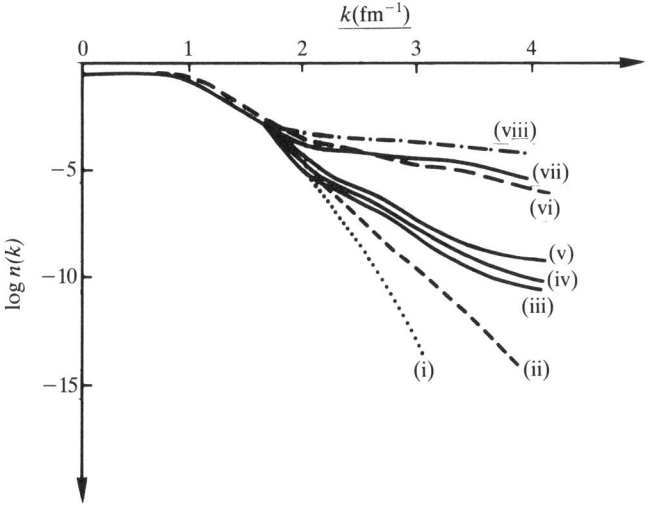

FIG. 7.7. The nucleon momentum distribution for ^{40}Ca. Curves correspond to: i) the harmonic-oscillator model (Traini and Orlandini 1985); ii) harmonic-oscillator with pairwise tensor correlations (Traini and Orlandini 1985); iii) Slater determinant total wavefunction of natural orbitals in the SPP method (Malaguti et al. 1982); iv) SPP with 3.55% of protons above the Fermi-level; v) SPP with 20.5% of protons above the Fermi-level; vi) CDFM (Antonov et al. 1980, 1983b); vii) harmonic oscillator with pairwise oscillator with pairwise tensor plus short-range correlations (Traini and Orlandini 1985); viii) calculations in the approach of Benhar et al. (1986). The curves (iii–v) are results of the SPP-method calculations of Antonov et al. (1987).

model (Antonov et al. 1980, 1983) (curve vi)) as well as the calculations in the approach of Benhar et al. (1986) (curve viii)). It is notable that the SPP method gives nucleon momentum distributions with much larger high-momentum components than the harmonic-oscillator model. This is partly attributable to the promotion of protons above the Fermi-sea, as shown by the differences between the curves iii)–v), but a much larger effect is due to the differences in the radial wavefunctions resulting from the requirement that they fit the density distributions. The curve iii) corresponds to the Slater-determinant total wavefunction constructed by SPP-method natural orbitals. This shows that the latter already includes much of the effects of short-range correlations.

The results of SPP calculations of the proton momentum distributions $n_p(k)/Z$ of ^{40}Ca are compared in Fig. 7.8 with several models of $n_p(k)/Z$ of ^{208}Pb considered by Jaminon et al. (1986). These models are: i) the Hartree–Fock theory with full occupancy of the proton single-particle

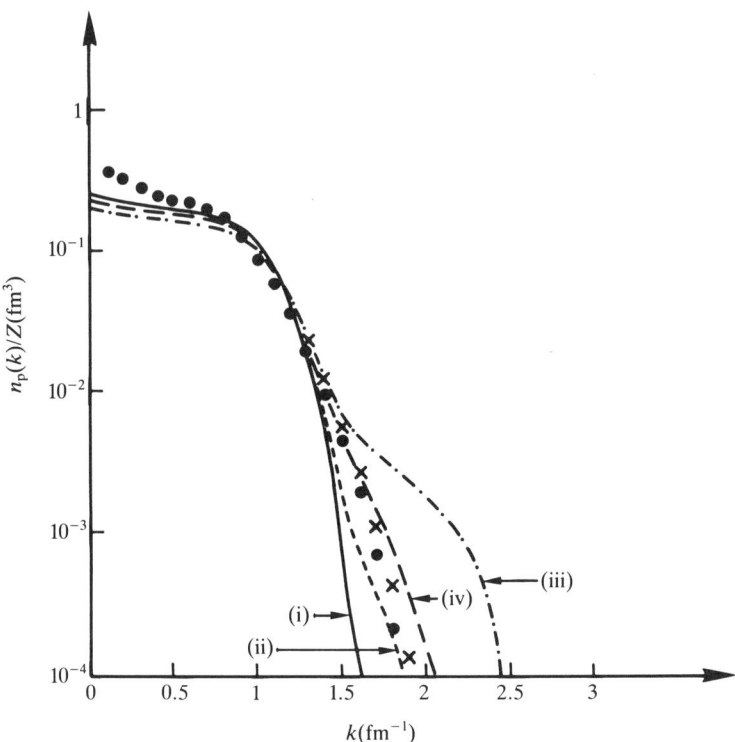

FIG. 7.8. The proton momentum distribution $n_p(k)/Z$ for ^{208}Pb calculated by Jaminon *et al.* (1986) (curves i)–iv)) compared with $n_p(k)/Z$ for ^{40}Ca calculated in the SPP method by Antonov *et al.* (1987). Curves correspond to: i) Hartree–Fock with simple shell-model occupation numbers; ii) RPA with 3.6% of protons above the Fermi-level; iii) RPA plus short-range and tensor correlations (20.6% of protons above the Fermi-level); iv) occupation numbers intermediate between cases ii) and iii) (11.6% of protons above the Fermi-level). The SPP-method calculations (Antonov *et al.* 1987) are made with 3.55% (solid dots) and 20.5% (crosses) of protons above the Fermi-level.

orbits up to the $3s_{\frac{1}{2}}$ level, with the higher level empty; ii) the random-phase approximation (RPA) with the occupation numbers of Dechargè and Sips (1983), with 3.6% of the protons above the Fermi-sea; iii) occupation numbers obtained by adding the effects of the RPA to those produced by short-range and tensor correlations (Pandharipande *et al.* (1984), with 20.6% of the protons above the Fermi sea; iv) occupation numbers intermediate between those of the last two models, with 11.6% of the protons above the Fermi-sea. The SPP calculations for ^{40}Ca were made with sets of occupation numbers corresponding to similar percentages of protons above the Fermi-sea. A comparison of these proton momentum distributions $n_p(k)/Z$ shows that the occupation numbers have relatively little effect on the SPP distribution which, as in ^{40}Ca, lies between the distributions corresponding to the Slater determinant with harmonic-oscillator wavefunctions (Hartree–Fock theory) and the theory including effects of tensor and short-range correlations.

The effect of the changed occupation numbers on the density distribution is shown in Fig. 7.9. It is notable that the major effects are in the interior of the nucleus, as was already found by Malaguti et al. (1982). The deviation of $\rho(r)$ near the origin for different occupation numbers correspond to the different behaviour of the momentum distribution at large momenta. As the results show, the tail of the momentum distribution is more sensitive to the particular form of the natural orbitals. We can conclude that the natural orbitals are responsible for an effective account of the short-range correlations.

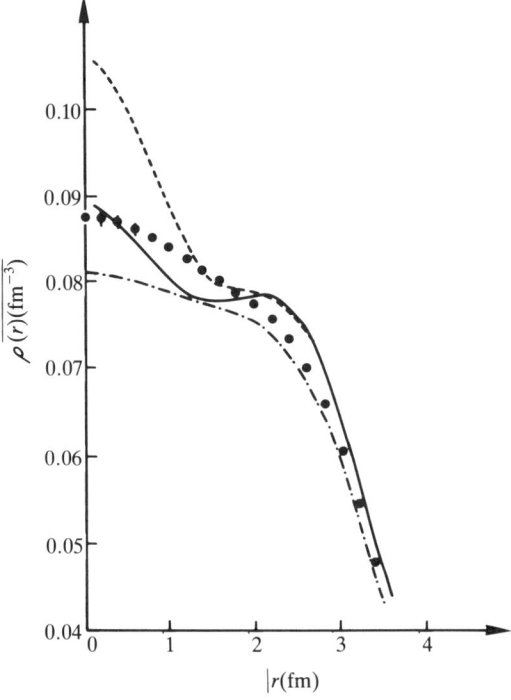

FIG. 7.9. Nuclear charge distribution of ^{40}Ca. (solid dots): experimental data (Sick I) taken from the paper of Malaguti et al. (1982); (dashes): calculations with Slater determinant wave-function of SPP natural orbitals; (solid line): SPP with 3.55% of protons above the Fermi-level; (dash-dot-line): SPP with 20.5% of protons above the Fermi-level. The theoretical curves are calculated by Antonov et al. (1987). The normalization is $4\pi \int_0^\infty dr\, r^2 \rho(r) = Z$.

7.4. Another phenomenological model

In the papers of Jaminon et al. (1985a, 1985b, 1986) a phenomenological attempt was made to construct in a rather elementary way 'natural orbitals'.

The first occupied single-particle wavefunctions are obtained from a Saxon–Woods potential with a particular depth-energy dependence. The parameters of the potentials are fitted so that the resulting energy-level scheme is close to that of self-consistent Hartree–Fock calculations (Decharge and Gogny 1980). The wavefunctions of the states above the Fermi-sea up to tens of MeV are constructed using a cut-off parameter R_0 and the boundary condition for the radial part of the wavefunction

$$\left[\frac{\mathrm{d}u_\alpha(r)}{\mathrm{d}r} \bigg/ u_\alpha(r) \right]_{r=R_0} = K, \quad K < 0. \tag{7.51}$$

In the region $r > R_0$ the wavefunction is of the form

$$u_\alpha(r) = u_\alpha(R_0) \exp[K(r - R_0)]. \tag{7.52}$$

This set of functions is assumed to be a set of natural orbitals.

In order to calculate observables such as density and momentum distributions the occupation numbers must be specified. In several studies of nuclear matter (for example (Flynn et al. 1984; Fantoni and Pandharipande 1984)) and finite systems (for example (Gaudin et al. 1971)) particular information for the occupied numbers and natural orbitals is already obtained. For nuclear matter the estimate of the number of particles above the Fermi-sea ($k > k_\mathrm{F}$) is about 20%. In finite nuclei the depletion of the Fermi-sea is estimated to be even more as was shown by interpretation of electron-scattering data (Frois et al. 1983; Pandharipande et al. 1984).

In the papers of Jaminon et al. (1985b, 1986) the occupation numbers are taken from Decharge and Sips (1983) computed in RPA and from nuclear-matter considerations, taking account of short-range and tensor correlations (Pandharipande et al. 1984) (PPW). The occupation numbers from RPA, PPW + RPA, and intermediate case numbers (PPWT + RPA) are shown in Fig. 7.10.

The result for the charge density:

$$\rho_\mathrm{c}(r) = (\pi a^2)^{-\frac{3}{2}} \int \rho_\mathrm{p}(r') \exp[-(r - r')^2/a^2] \, \mathrm{d}r', \tag{7.53}$$

with $a^2 = 0.4 \, \mathrm{fm}^2$, and using PPWT + RPA occupation numbers is illustrated in Fig. 7.11 compared with the Hartree–Fock charge density (all occupation numbers $n_\alpha = 1$ at $\alpha \leq \alpha_\mathrm{F}$ and $n_\alpha = 0$ at $\alpha > \alpha_\mathrm{F}$) and the experimental density (Frois et al. 1977).

NATURAL-ORBITAL CALCULATIONS

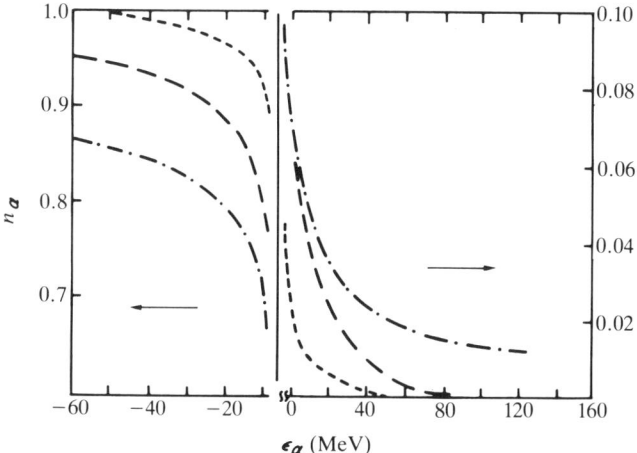

FIG. 7.10. Dependence of the proton occupation probabilities $n_\alpha(n, l, j)$ upon the single-particle energy ε_α, in the case of RPA (short dashes); PPW + RPA (dash dot); and PPWT + RPA (long dashes) models for the ground state of ^{208}Pb (Jaminon et al. 1985b, 1986).

The nucleon momentum distribution

$$n(\boldsymbol{k}) = \sum n_\alpha \tilde{\varphi}_\alpha^*(\boldsymbol{k}) \tilde{\varphi}_\alpha(\boldsymbol{k}), \tag{7.54}$$

$$\int n(\boldsymbol{k}) \, \mathrm{d}\boldsymbol{k} = Z, \tag{7.55}$$

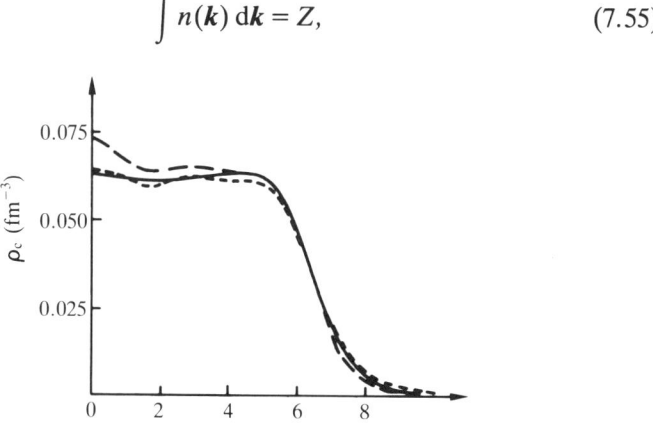

FIG. 7.11. Charge-density distribution in ^{208}Pb (Jaminon et al. 1985b). The solid line represents the experimental density (Frois et al. 1983). The long dashes represent the Hartree–Fock-type calculations of Jaminon et al. (1985b). The short dashes show the effect of a depletion of the PPWT + RPA type.

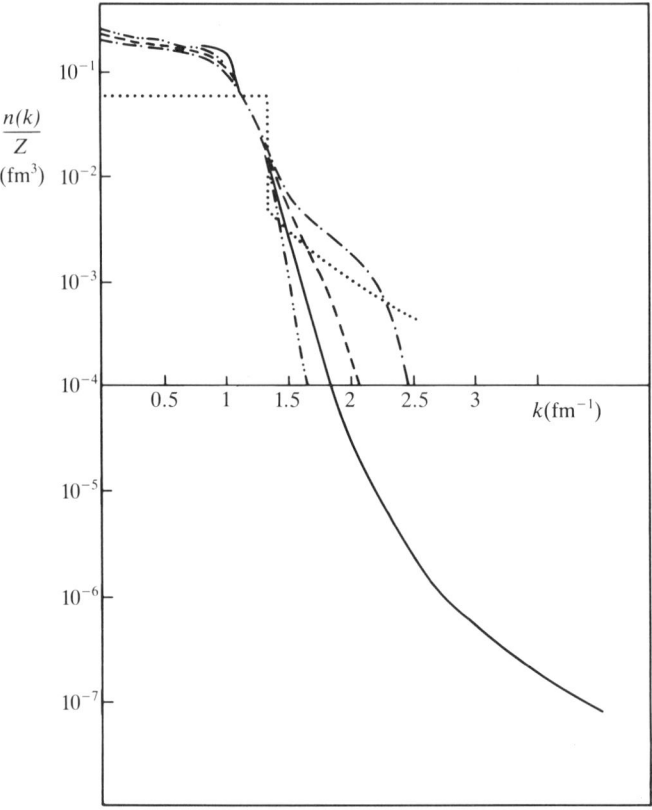

FIG. 7.12. Proton momentum distribution in ^{208}Pb from the calculation in Hartree–Fock (dash two dots); PPW + RPA (dash dot); PPWT + RPA (dashed line) approaches (Jaminon et al. 1986), from the CDFM calculation (Antonov and Petkov 1987) (solid line). The dotted curve represents the proton momentum distribution in nuclear matter with $k_F = 1.33$ fm^{-1} (Fantoni and Pandharipande 1984).

where $\tilde{\varphi}_\alpha(k)$ are the Fourier transforms of the natural orbitals $\varphi_\alpha(r)$, calculated with PPWT + RPA occupation numbers is presented in Fig. 7.12 and compared with Hartree–Fock calculations, with the results of $n(k)$ derived from the CDFM (Antonov and Petkov 1987) and nuclear-matter calculations for $n(k)$ with $k_F = 1.33$ fm^{-1} (Fantoni and Pandharipande 1984).

The results for the density and momentum distributions given in Figs. 7.11 and 7.12 show the role of occupation numbers $n_\alpha < 1$, meaning that some states are partially unoccupied ($\alpha < \alpha_F$) and partially occupied ($\alpha > \alpha_F$) or, alternatively, demonstrate the Fermi-sea depletion due to the short-range and tensor correlations. The main effect of the depletion on the density distributions is that it becomes flatter in the inner region, in better agreement with the experimental data. As for the momentum distribution the depletion causes the rise of the distribution in the domain $1.8 \text{ fm}^{-1} \le k \le 2.2 \text{ fm}^{-1}$ which is in accord with other more sophisticated theoretical approaches incorporating short-range correlations (Zabolitzky and Ey 1978; Van Orden *et al.* 1980).

It is seen from Fig. 7.11 that the result for the density distribution calculated from the Slater determinant composed of natural orbitals does not reproduce the experimental density. Let us note, however, that the 'exact' density distribution for ^{208}Pb calculated with natural orbitals deviates from the density calculated with the Slater determinant of the same natural orbitals. This result contradicts the calculation of $\rho(r)$ in ^{40}Ca (Gaudin *et al.* 1971) using natural orbitals resulting from diagonalization of a Jastrow one-body density matrix in comparison with the density calculated with a Slater determinant constructed with the same natural orbitals. In the case of ^{40}Ca these two calculated densities are in good agreement.

We note that the reason for the contradiction between the cases of ^{208}Pb and ^{40}Ca probably comes from the fact that in the former case the occupation numbers are not consistent with the natural orbitals used as well as from the approximate nature of their functional form.

8

OTHER PHENOMENOLOGICAL MODELS. EXPERIMENTAL DATA RELATED TO THE NUCLEON MOMENTUM DISTRIBUTION

In Section 8.1 of this chapter phenomenological models which use Wigner-distribution functions for analyses of the momentum distribution are given. The generator-coordinate method with Skyrme-type forces and different construction potentials is applied to the description of both the density and momentum distributions (Section 8.2). A phenomenological model accounting for the nuclear finite-size effects is described in Section 8.3. The problems of the determination of the nucleon momentum distribution from the experimental data are discussed in Section 8.4.

8.1. Finite-size effects in the nucleon momentum distribution

The mean-field calculations of the nucleon momentum distribution exhibit two characteristic regions: i) an outer region ($k > 1 \, \text{fm}^{-1}$) where the slope of $n(k)$ is nearly common to all nuclei, i.e. for which the finite-size effects are not essential, and ii) an interior region ($k < 1 \, \text{fm}^{-1}$) which is different for different nuclei, i.e. it is more strongly influenced by nuclear finite-size effects. This situation is well illustrated in Fig. 8.1 where the density-dependent Hartree–Fock (DDHF) calculations of $n(k)$ are presented for ^{208}Pb, ^{120}Sn, and ^{90}Zr (Casas *et al.* 1986). For comparison, the $n(k)$ for a non-interacting Fermi-gas is presented in the same figure. As can be seen there is a striking difference between the nucleon momentum distribution in finite nuclei and in the infinite system. The physical reason for this is that in the infinite system the momentum k is a good quantum number while in finite systems k is a quantum number only with an accuracy $\Delta k \sim 1/R$, where R is the size of the system. Therefore even in the case of heavy nuclei the finite-size effects are of importance.

Quite different is the situation of the local density distribution in finite systems and in nuclear matter. It is well established that the internal region density for nuclei (especially for heavy ones) is almost constant and close to that of infinite nuclear matter.

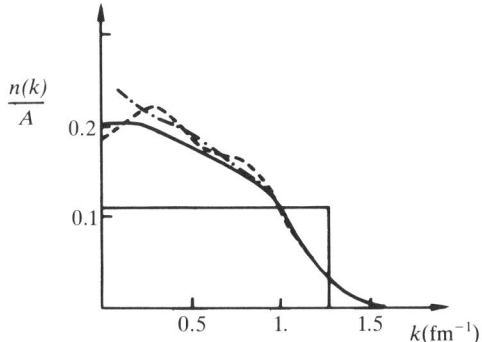

FIG. 8.1. Momentum distributions from DDHF calculations of Casas *et al.* (1986) for ^{208}Pb (solid line); ^{120}Sn (dash-dotted line) and ^{90}Zr (dashed line) in comparison with the nuclear-matter result (step function).

It was shown in Section 2.1 that these two main nuclear characteristics, $\rho(r)$ and $n(k)$ are connected by the Wigner-distribution function $W(r, k)$:

$$\rho(r) = \int W(r, k) \frac{dk}{(2\pi)^3}, \tag{8.1}$$

$$n(k) = \int W(r, k) \, dr. \tag{8.2}$$

For infinite nuclear matter with density ρ_0 the Wigner-distribution function has the simple form:

$$W(r, k) = 4\theta(k_F(\rho_0) - k), \tag{8.3}$$

where

$$k_F(\rho_0) = \left(\frac{3\pi^2}{2} \rho_0\right)^{\frac{1}{3}}, \tag{8.4}$$

and the relations (8.1) and (8.2) lead to $\rho = \rho_0$ and $n(k) = 4\Omega\theta(k_F(\rho_0) - k)$.

On the basis of the Wigner function (8.3) and replacing $k_F(\rho_0)$ with $k_F(\rho(r))$ Hüfner and Nemes (1981) developed the local Fermi-gas approximation for finite nuclei. In this case the proton momentum distribution is of the form

$$n(k) = 8\pi \int \theta(k_F(\rho(r)) - k) r^2 \, dr, \tag{8.5}$$

with

$$k_F(\rho(r)) = (3\pi^2 \rho(r))^{\frac{1}{3}}. \tag{8.6}$$

Two main features of this momentum distribution can be noticed: i) $n(k)$ diverges when $k \to 0$ and if $\rho_p(r)$ decreases exponentially for large values of r, then $n(k)$ diverges like $(\ln k)^3$ when $k \to 0$, ii) $n(k)$ vanishes for $k > k_{\max} = (3\pi^2 \rho_{\max})^{\frac{1}{3}}$, were $\rho_{\max}$ is the maximum value taken by the proton density distribution.

It was noticed, however, in (Jaminon et al. 1986) that in the domain $0.5 < k < 1 \text{ fm}^{-1}$ the local Fermi-gas approach is approximately in good agreement with the Hartree–Fock results.

Similar analyses have been carried out by Casas et al. (1986) using different forms of the local density distribution (trapezoidal density, cubic trapezoidal density, and diffuse Fermi-density). The calculations show that $n(k)$ is in good agreement with DDHF results in the domain $0.5 < k < 1 \text{ fm}^{-1}$. An effort is made to overcome the shortcomings of the model at $k \to 0$ and $k > k_{\max}$ using a Wigner function which is the step function

$$W(\mathbf{r}, \mathbf{k}) = 4\theta(k_F(\rho(r)) - k), \tag{8.7}$$

convoluted with a Gaussian function

$$W_C(\mathbf{r}, \mathbf{k}) = \int W(\mathbf{r}, \mathbf{k}')g(|\mathbf{k} - \mathbf{k}'|)\,\mathrm{d}\mathbf{k}', \tag{8.8}$$

where

$$g(|\mathbf{k} - \mathbf{k}'|) = (\pi\mu^2)^{-\frac{3}{2}} \exp(-|\mathbf{k} - \mathbf{k}'|^2/\mu^2), \tag{8.9}$$

and μ is the free parameter.

It is interesting to note that in this simple method the DDHF results are reproduced rather well as can be seen in Fig. 8.2 in the case of ^{208}Pb, apart from the small region of $k \leq 0.5 \text{ fm}^{-1}$.

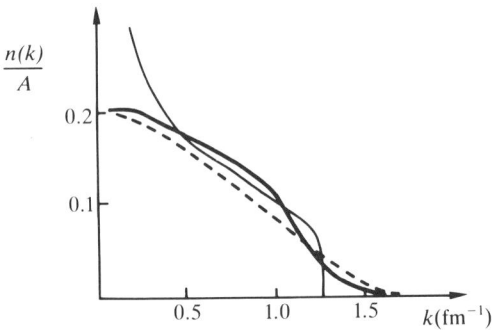

FIG. 8.2. Results for $n(k)/A$ for ^{208}Pb (Casas et al. 1986) obtained from different approximations to the Wigner distribution function: eqn (8.7) (thin solid line); eqns (8.8) and (8.9) with $\mu = 0.4$ (dashed line) in comparison with the results from the DDHF calculation (thick solid line).

An analysis of the internal region of the momentum distribution in finite nuclei has been made by Zverev and Saperstein (1986) using the Lagrangian quasiparticle method (Khodel and Saperstein 1982). The Wigner wavefunction is used in the form:

$$W(r, k) = \int \frac{d\varepsilon}{2\pi i} \int ds\, e^{ip\cdot s} G(r + \tfrac{1}{2}s, r - \tfrac{1}{2}s; \varepsilon), \tag{8.10}$$

where the one-particle Green function $G(\varepsilon)$ in the finite Fermi system (Migdal 1967) is expressed as a sum of the regular component $G_R(\varepsilon)$ and the quasiparticle Green function $G_q(\varepsilon) = (\varepsilon - \varepsilon_p^0 - \Sigma_q)^{-1}$, where $\Sigma_q(r, k, \varepsilon)$ is the quasiparticle mass operator.

Correspondingly, the momentum distribution is expressed in the form:

$$n(k) = v(k) + n_R(k), \tag{8.11}$$

where the quasiparticle momentum distribution $v(k)$ is generated by G_q and $n_R(k)$ by G_R.

In eqn (8.11)

$$v(k) = \sum_\lambda n_\lambda |\psi_\lambda(k)|^2, \tag{8.12}$$

where the n_λ are the quasiparticle occupation numbers,

$$\psi_\lambda(k) = \int dr\, e^{ik\cdot r} \psi_\lambda(r), \tag{8.13}$$

and $\psi_\lambda(r)$ are functions diagonalizing $G_q(\varepsilon)$. The authors Zverev and Saperstein (1986) accepted a model in which the nucleon momentum distribution consists of a quasiparticle part $v(k)$ calculated in the Lagrangian quasiparticle method for the finite nucleus and a regular tail believed to be insensitive to the finite-size effects, $n_R(k)$ taken from nuclear-matter calculations. These two distributions are smoothly connected and the result is properly normalized to the number of particles. The calculations for a number of nuclei ^{208}Pb, ^{132}Sn, ^{90}Zr, ^{48}Ca, and ^{40}Ca have been carried out and the results for heavier nuclei are in overall agreement with the DDHF method in the interior region of k ($k < 1$ fm^{-1}).

Finite-size effects on the nucleon momentum distribution of non-interacting fermions are considered also by Krivine (1986). The momentum distributions of A non-interacting fermions enclosed in a one-dimensional infinite potential well, as well as in a three-dimensional cubic, parallelepipedic, and spherical domains are obtained. For example, the momentum distribution of fermions in a sphere is of the form:

$$\frac{n(k)}{n_0} = \frac{6}{\pi^2} \sum_{l,p_l} (2l+1) j_l^2(kR) \frac{(k_{p_l}R/\pi)^2}{[(k_{p_l}R/\pi)^2 - (kR/\pi)^2]^2}, \tag{8.14}$$

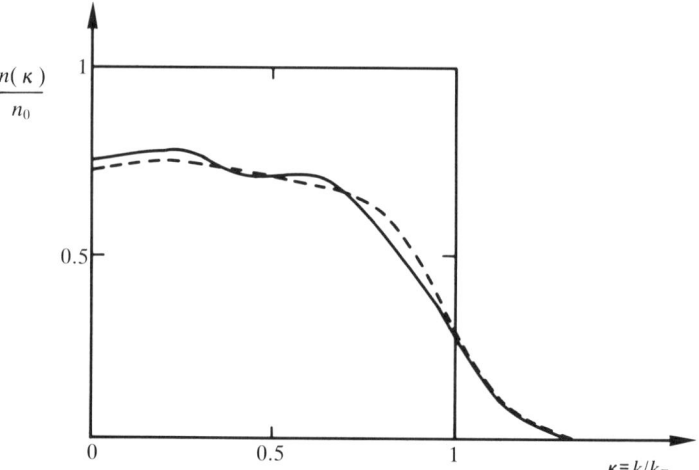

FIG. 8.3. Comparison between the momentum distribution of $A = 136$ nucleons enclosed in a cubic domain (dashed line) and in a sphere (solid line). The results are taken from the paper of Krivine (1986).

where R is the radius of the sphere, j_l is the spherical Bessel function, k_{p_l} is such that $j_l[k_{p_l}R] = 0$.

In (8.14)

$$n_0 = vV/(2\pi\hbar)^3, \tag{8.15}$$

where v is the spin-isospin degeneracy and V is the volume of the domain. The momentum distribution (8.14) is shown in Fig. 8.3 and compared with that one in the case of a cubic domain.

The effects of the finiteness of the potential wells are also examined in the work of Krivine (1986). The changes of $n(k)$ generated by the finite size of the system are compared to those due to the finiteness of the nucleons, and it is concluded that in the nuclear case both effects are of the same order of magnitude.

8.2. An approach to nucleon momentum and density distributions in the generator-coordinate method

In the generator-coordinate method (GCM) (Hill and Wheeler 1953; Griffin and Wheeler 1957; Wong 1975) the trial many-body wave-function is written in the form of the linear combinations

$$\psi(r_1, \ldots r_A) = \int f(x_1, x_2, \ldots)\Phi(r_1, \ldots, r_A; x_1, x_2, \ldots) \, dx_1 \, dx_2 \ldots \tag{8.16}$$

The function $\Phi(\{r_i\}; x_1, x_2, \ldots)$, $(i = 1, 2, \ldots A)$ is called the generating function. It depends on the particle coordinates $\{r_i\}$ and on the generator coordinates $x_1, x_2, \ldots$. The unknown function $f(x_1, x_2, \ldots)$ of the generator coordinates is called the generator or weight function.

Application of the Ritz variational principle $\delta E = 0$ leads to the integral equation for the weight function:

$$\int [\mathcal{H}(x, x') - EI(x, x')]f(x') \, dx' = 0, \tag{8.17}$$

where for simplicity we use only one generator coordinate x. The kernels in (8.17) are of the forms:

$$I(x, x') = \langle \Phi(\{r_i\}, x) \mid \Phi(\{r_i\}, x') \rangle, \tag{8.18}$$

$$\mathcal{H}(x, x') = \langle \Phi(\{r_i\}, x) \mid \hat{H} \mid \Phi(\{r_i\}, x') \rangle. \tag{8.19}$$

The operator

$$\hat{H} = \sum_i \frac{p_i^2}{2m} + \sum_{i<j} V_{ij} \tag{8.20}$$

is the Hamiltonian of the system considered.

The solutions f_i, f_j of eqn (8.17) corresponding to the energy eigenvalues E_i, E_j satisfy the orthonormality condition:

$$\int f_i^*(x) I(x, x') f_j(x') \, dx \, dx' = \delta_{ij}. \tag{8.21}$$

Eqn (8.17) is called the Griffin–Hill–Wheeler (GHW) generator-coordinate wave equation.

In the case of a large number of nucleons the kernels (8.18) and (8.19) are strongly peaked at $x \approx x'$ and can be written (Wildermuth and Tang 1977):

$$I(x, x') \simeq I(x, x) F(x - x'), \tag{8.22}$$

$$\mathcal{H}(x, x') \simeq \mathcal{H}(x, x) F(x - x'), \tag{8.23}$$

where the function $F(x - x')$ is peaked at $x \simeq x'$. This property allows us to use the expansion of the weight function:

$$f(x') = f(x) + \frac{df(x)}{dx}(x' - x) + \frac{1}{2!}\frac{d^2 f(x)}{dx^2}(x' - x)^2 + \ldots \tag{8.24}$$

Substituting (8.24) in (8.17) and accounting for (8.22) and (8.23) one obtains the Schrödinger equation:

$$-\frac{\hbar^2}{2m_{\text{eff}}(x, E)} \frac{d^2 f(x)}{dx^2} + V(x) f(x) = E f(x). \tag{8.25}$$

Here

$$V(x) = \langle \Phi(\{r_i\}, x) | \hat{H} | \Phi(\{r_i\}, x) \rangle, \tag{8.26}$$

$$m_{\text{eff}}(x, E) = -\hbar^2 \left[\frac{1}{K(x)} \int \langle \Phi(\{r_i\}, x) | \hat{H} - E | \Phi(\{r_i\}, x') \rangle \right.$$
$$\left. \times (x' - x)^2 \, dx' \right]^{-1}, \tag{8.27}$$

$$K(x) = \int F(x' - x) \, dx'. \tag{8.28}$$

In practice the many-particle generating function $\Phi(\{r_i\}, x)$ is taken to be a Slater determinant constructed from neutron and proton orbitals $\varphi_\lambda(r, x)$. In this case the GCM in principle goes beyond the Hartree–Fock method and the extent of the correlations which can be involved depends on the number and nature of the generator coordinates.

In the case of one generator coordinate the density and momentum distributions can be examined simultaneously using different generating functions.

In the work of Antonov et al. (1986a) the generating wave-function is taken to be a Slater determinant $\Phi(\{r_i\}, x)$ built from neutron and proton orbitals $\varphi_\lambda(r, x)$ in a 'construction potential' of a square well with infinite walls. The radius of the well x is chosen as the generator coordinate. At the same time it is related to the size of the system of A nucleons confined in the sphere with radius x; i.e. it is the radius of the system in the 'intermediate' state $\Phi(\{r_i\}, x)$. The dynamics related to this collective parameter is the breathing-monopole nuclear vibration.

The Skyrme forces are used for which the Hamiltonian kernel (8.19) of the GHW equation is factorized (Brink 1966):

$$\mathcal{H}(x, x') = \langle \Phi(\{r_i\}, x) | \Phi(\{r_i\}, x') \rangle \int H(x, x', r) \, dr. \tag{8.29}$$

For $N = Z$, $H(x, x', r)$ takes the form (Flocard and Vautherin 1975, 1976):

$$H(x, x', r) = \frac{\hbar^2}{2m} T + \tfrac{3}{8} t_0 \rho^2 + \tfrac{1}{16}(3t_1 + 5t_2)(\rho T + j^2)$$
$$+ \tfrac{1}{64}(9t_1 - 5t_2)(\nabla \rho)^2 + \tfrac{1}{16} t_3 \rho^{2+\sigma}. \tag{8.30}$$

The quantities t_0, t_1, t_2, t_3, and σ are Skyrme force parameters. The density ρ, the kinetic-energy density T, and the current density j are defined by

$$\rho(x, x', r) = 4 \sum_{\lambda, \mu = 1}^{A/4} (N^{-1})_{\mu\lambda} \varphi_\lambda^*(r, x) \varphi_\mu(r, x'), \tag{8.31}$$

$$T(x, x', r) = 4 \sum_{\lambda,\mu=1}^{A/4} (N^{-1})_{\mu\lambda} \nabla \varphi_\lambda^*(r, x) \nabla \varphi_\mu(r, x'), \qquad (8.32)$$

$$j(x, x', r) = 2 \sum_{\lambda,\mu=1}^{A/4} (N^{-1})_{\mu\lambda} \{\varphi_\lambda^*(r, x) \nabla \varphi_\mu(r, x') - [\nabla \varphi_\lambda^*(r, x)] \varphi_\mu(r, x')\},$$
$$(8.33)$$

where A is the mass number of the nucleus and

$$N_{\lambda\mu} = \int \varphi_\lambda^*(r, x) \varphi_\mu(r, x') \, dr. \qquad (8.34)$$

The overlap kernel (8.22) is given by:

$$I(x, x') = [\det(N_{\lambda\mu})]^4. \qquad (8.35)$$

The nucleon density distribution $\rho(r)$ and the nucleon momentum distribution $n(k)$ are expressed in terms of the weight function $f_0(x)$ as follows:

$$\rho(r) = \int f_0(x) f_0(x') I(x, x') \rho(x, x', r) \, dx \, dx', \qquad (8.36)$$

$$n(k) = \int f_0(x) f_0(x') I(x, x') \rho(x, x', k) \, dx \, dx', \qquad (8.37)$$

where

$$\rho(x, x', k) = 4 \sum_{\lambda,\mu=1}^{A/4} (N^{-1})_{\mu\lambda} \tilde{\varphi}_\lambda^*(k, x) \tilde{\varphi}_\mu(k, x'). \qquad (8.38)$$

The function $f_0(x)$ is the lowest solution of the GHW equation (8.17) corresponding to the ground-state energy. The function $\tilde{\varphi}_\lambda(k, x)$ is the Fourier transform of $\varphi_\lambda(r, x)$.

The GCM results for ^{4}He and ^{16}O nuclei are shown in Figs. 8.4–8.7. In solving eqn (8.17) Skyrme-like forces are used.

The values of the Skyrme parameters fixed in order to reproduce the binding energy of the ^{16}O nucleus are: $t_0 = -2765$, $t_1 = 317.33$, $t_2 = -38.04$, $t_3 = 16263$, and $\sigma = \frac{1}{6}$. The nuclear-matter characteristics corresponding to this set of Skyrme parameters are $E/A = -15.98$ MeV/A, the compressibility $K = 221.1$ MeV, $m^*/m = 0.746$, and the equilibrium density $\rho_0 = 0.148$ fm^{-3}.

The proton density distributions of ^{4}He and ^{16}O obtained in this way are compared in Figs. 8.4 and 8.5 with those from the GCM approach with harmonic-oscillator functions and Hartree–Fock calculations (both with SkM* forces (Krivine et al. 1980)). It can be seen that in spite of the fact that the generating functions are non-zero only in a finite volume, the resulting density distributions have the correct asymptotic form. This

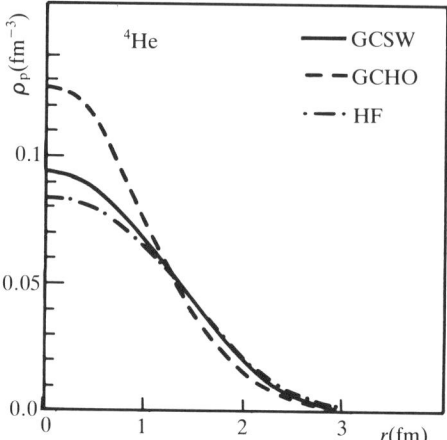

FIG. 8.4. Proton density distribution of ^{4}He (Antonov *et al.* 1986*a*) calculated with square-well functions in the generator-coordinate method (GCM) (solid line), with harmonic-oscillator functions in GCM and SkM* forces (dashed line) and with Hartree–Fock methods (dash-dotted line).

is apparently due to the account of the collective zero-motion dynamics in the GCM procedure. The evaluated densities are close to the Hartree–Fock ones in the important surface region for both nuclei. There is a difference only in the central region of the ^{16}O density.

In Figs. 8.6 and 8.7 the results for the nucleon momentum distributions of ^{4}He and ^{16}O are compared with those from the exp(S) method (Zabolitzky and Ey 1978) and also with the CDFM calculation (Antonov

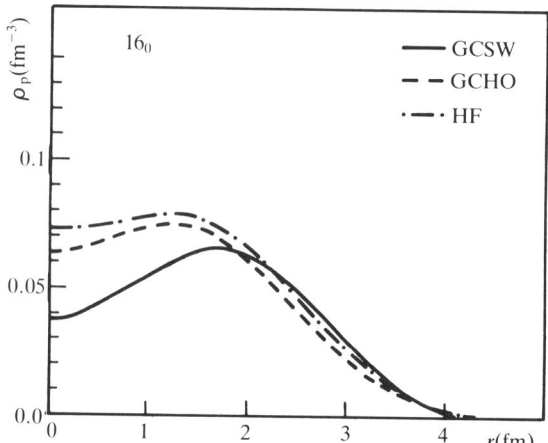

FIG. 8.5. Same as in Fig. 8.4 for ^{16}O (Antonov *et al.* 1986*a*).

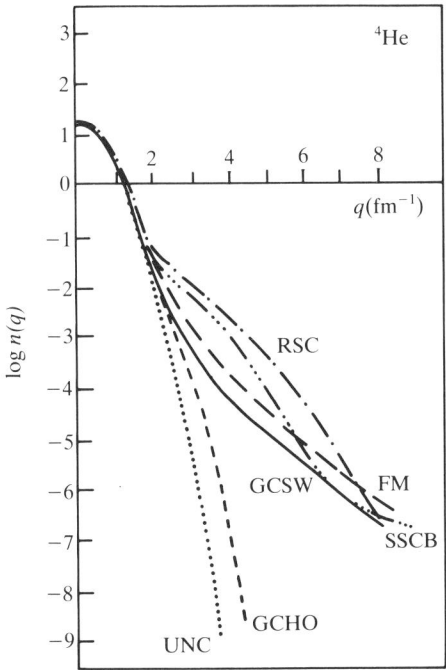

FIG. 8.6. Nucleon momentum distribution for ^{4}He (Antonov et al. 1986a) calculated in GCM with square-well single-particle functions (GCSW) and in GCM with harmonic-oscillator functions and SkM* forces (GCHO). The exp(S) calculations (Zabolitzky and Ey 1978) use different potentials: Reid soft core (RSC); de Tourreil–Sprung super soft core B (SSCB); uncorrelated for Reid soft core (UNC); and from the CDFM (FM) (Antonov et al. 1979, 1980).

et al. 1979, 1980). The momentum distributions calculated with square-well functions are close to those of (Zabolitzky and Ey 1978; Antonov et al. 1980) and exhibit a high-momentum component at $k > 2$ fm^{-1}. On the other hand they differ significantly from both the momentum distributions calculated with GCM, using harmonic-oscillator functions, and the case of uncorrelated nucleons. It turns out that the behaviour of the momentum distributions does not depend essentially on the particular choice of Skyrme parameter set (SkIII (Beiner et al. 1975) or SkM*) in the square-well GCM calculations. As shown, for momenta $k > 2$ fm^{-1} the momentum distribution is strongly affected by short-range correlations. Obviously, the inclusion of only the monopole vibrations is not sufficient to account for all the effects of the short-range correlations, as is illustrated by the GCM harmonic-oscillator results. Therefore, the choice of the intermediate generating states is very important for a

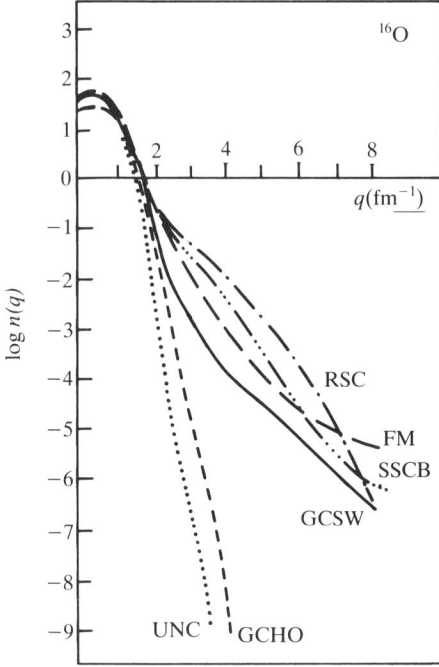

FIG. 8.7. Same as in Fig. 8.6 for ^{16}O (Antonov et al. 1986a).

correct account of the effects of short-range correlations. It could be stated that this approach, which assumes that the nucleons are confined to a finite volume in the intermediate states, accounts rather effectively for part of the short-range correlations.

The functions $\rho(x, x, k)$ (8.38) for square-well generating and harmonic-oscillator functions together with the corresponding momentum distributions $n(k)$ are plotted in Fig. 8.8. The behaviour of $\rho(x, x, k)$ and $n(k)$ in the case of square-well functions illustrates that the collective monopole vibrations are also important.

In Table 8.1 the energy values and r.m.s. radii of the ground and first excited monopole states of ^{4}He and ^{16}O are presented.

Obviously, the use of a square-well potential with infinite walls for studying monopole vibrations in the GCM leads to: i) the presence of a high-momentum component at $k > 2\,\text{fm}^{-1}$ in the nucleon momentum distribution, which provides evidence of some types of short-range correlations, and ii) a correct behaviour of the local density distribution in the important nuclear surface region.

OTHER PHENOMENOLOGICAL MODELS

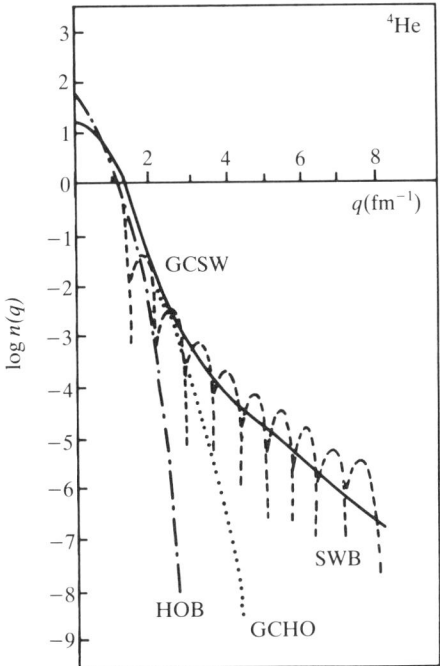

FIG. 8.8. The function $\rho(r, r, q)$ (8.38) and the nucleon momentum distribution $n(q)$ (8.37) for ^{4}He (Antonov *et al.* 1986a). The quantity r is the r.m.s. radius in the chosen intermediate state (in this example $r = 2.6$ fm). SWB: $\rho(r, r, q)$ in the generator-coordinate method (GCM) with square-well generating functions (dashed line); GCSW: $n(q)$ in GCM with square-well generating functions (solid line); HOB: $\rho(r, r, q)$ in GCM with harmonic-oscillator generating functions (dash-dotted line); GCHO: $n(q)$ in GCM with harmonic-oscillator generating functions (dotted line).

TABLE 8.1. Energies (in MeV) and r.m.s. radii (in fm) of the ground and first excited monopole states of ^{4}He and ^{16}O obtained by the generator-coordinate method, with an infinite square-well potential (Antonov *et al.* 1986a).

Nucleus	E_0	r_0	E_1	r_1	ΔE_1
^{4}He	−36.54	1.758	−9.6	2.80	26.94
^{16}O	−139.96	2.62	−107.28	2.89	32.68

8.3. The coherent density fluctuation model

The differential equation (8.25) in the GCM for the weight function $f(x)$ can be formally obtained using the so-called delta-function approximation (Griffin and Wheeler 1957) for the kernels $I(x, x')$ (8.18) and $\mathcal{H}(x, x')$ 8.(19):

$$I(x, x') \to \delta(x - x'), \tag{8.39}$$

$$\mathcal{H}(x, x') \to -\frac{\hbar^2}{2m_{\text{eff}}} \delta''(x - x') + \delta(x - x') V\left(\frac{x + x'}{2}\right). \tag{8.40}$$

If the trial many-body wavefunction in GCM:

$$\psi(\{r_i\}) = \int_0^\infty f(x) \Phi(x, \{r_i\}) \, dx, \quad i = 1, 2, \ldots A, \tag{8.41}$$

is normalized to the mass number A and the weight function satisfies the normalization condition:

$$\int_0^\infty |f(x)|^2 \, dx = 1, \tag{8.42}$$

then using (8.39) $I(x, x')$ becomes:

$$\int \Phi^*(\{r_i\}, x') \Phi(\{r_i\}, x) \, dr_1 \ldots dr_A = A \delta(x - x'). \tag{8.43}$$

By analogy with the relation (8.43) one can hope that in the case of a many-particle system the following relation will approximately hold:

$$\int \Phi^*(r, r_2, \ldots r_A; x') \Phi(r', r_2, \ldots r_A; x) \, dr_2 \ldots dr_A = \rho_x(r, r') \delta(x - x'), \tag{8.44}$$

where $\rho_x(r, r')$ is the one-body density matrix corresponding to the system described by the function $\Phi(\{r_i\}, x)$:

$$\rho_x(r, r') = \int \Phi^*(r, r_2, \ldots r_A; x) \Phi(r', r_2, \ldots r_A; x) \, dr_2 \ldots dr_A. \tag{8.45}$$

Let us assume that the function $\Phi(\{r_i\}, x)$ corresponds to the state of the system with a uniform density $\rho_x(r) \equiv \rho_x(r, r' = r)$:

$$\rho_x(r) = \rho_0(x) \theta(x - |r|), \tag{8.46}$$

$$\rho_0(x) = 3A/4\pi x^3. \tag{8.47}$$

The generator coordinate x corresponds here to the radius of a sphere which contains all A nucleons uniformly distributed in it.

OTHER PHENOMENOLOGICAL MODELS

The nuclear-matter density matrix is appropriate for description of such an object. It can be written in the form:

$$\rho_x(\mathbf{r}, \mathbf{r}') = 3\rho_0(x) \frac{j_1(k_F(x)|\mathbf{r}-\mathbf{r}'|)}{k_F(x)|\mathbf{r}-\mathbf{r}'|} \theta(x - \tfrac{1}{2}|\mathbf{r}+\mathbf{r}'|), \quad (8.48)$$

and its diagonal elements obviously coincide with the expression (8.46).

Now the one-body density matrix is completely determined as a coherent superposition of the one-body matrices for samples of nuclear matter with different densities $\rho_0(x)$:

$$\rho(\mathbf{r}, \mathbf{r}') = \int_0^\infty |f(x)|^2 \rho_x(\mathbf{r}, \mathbf{r}') \, dx. \quad (8.49)$$

The Wigner distribution function:

$$W(\mathbf{k}, \mathbf{r}) = \int \rho(\mathbf{r}+\tfrac{1}{2}\boldsymbol{\eta}, \mathbf{r}-\tfrac{1}{2}\boldsymbol{\eta}) e^{-i\mathbf{k}\cdot\boldsymbol{\eta}} \, d\boldsymbol{\eta} \quad (8.50)$$

can now be written in the explicit form:

$$W(\mathbf{k}, \mathbf{r}) = \int_0^\infty |f(x)|^2 \theta(x-|\mathbf{r}|)\theta(k_F(x)-|\mathbf{k}|) \, dx. \quad (8.51)$$

By means of eqn (8.51) it is easy to determine the nucleon density distribution:

$$\rho(\mathbf{r}) = 4 \int W(\mathbf{k}, \mathbf{r}) \frac{d\mathbf{k}}{(2\pi)^3}, \quad (8.52)$$

and the nucleon momentum distribution

$$n(\mathbf{k}) = \int W(\mathbf{k}, \mathbf{r}) \, d\mathbf{r}, \quad (8.53)$$

in their explicit form (Antonov et al. 1979, 1980):

$$\rho(r) = \int_0^\infty |f(x)|^2 \rho_0(x) \theta(x-r) \, dx, \quad (8.54)$$

$$n(k) = \int_0^\infty |f(x)|^2 \tfrac{24}{3}\pi x^3 \theta(k_F(x)-k) \, dx. \quad (8.55)$$

The relation (8.54) can be used for the determination of the function $|f(x)|^2$:

$$|f(x)|^2 = -\frac{1}{\rho_0(x)} \frac{d\rho(r)}{dr}\bigg|_{r=x}. \quad (8.56)$$

It is important to note that this relation holds only for monotonically-decreasing density distributions ($d\rho/dr < 0$). This is related to the delta-function limit used in GCM.

As a result the nucleon momentum distribution can be obtained using (8.55) and (8.56) as a functional of the density ρ (Antonov et al. 1979, 1980; Antonov and Petkov 1986);

$$n(k) = \left(\frac{4\pi}{3}\right)^2 \frac{4}{A} \left[6 \int_0^{\alpha/k} \rho(x) x^5 \, dx - \left(\frac{\alpha}{k}\right)^6 \rho\left(\frac{\alpha}{k}\right)\right], \quad (8.57)$$

where $\alpha \equiv (9\pi A/8)^{\frac{1}{3}}$, with the normalization condition

$$\int n(k) \frac{dk}{(2\pi)^3} = A. \quad (8.58)$$

Some general properties of the expression for $n(k)$ eqn (8.57) are:
i) the power-law fall-off of $n(k)$:

$$n(k) \sim k^{-8} \quad \text{as } k \to \infty, \quad (8.59)$$

under the conditions that

$$d\rho/dr|_{r=0} = 0, \quad d^2\rho/dr^2|_{r=0} \neq 0, \quad (8.60)$$

which is qualitatively in accord with the estimate from (Amado 1976) (see Section 3.3).

ii) As $k \to 0$:

$$n(k \to 0)/A \simeq (128\pi^2/3A^2) \int_0^\infty \rho(x) x^5 \, dx. \quad (8.61)$$

This expression is almost independent of A and its value is twice that in nuclear matter ($n(k \to 0)/A = 4/\rho_0$). This result is in accord with other estimates (Schiavilla et al. 1986; Casas et al. 1986).

iii) The expression for $n(k)$ eqn (8.57) is easily applicable to all nuclei with monotonic density distributions.

Some particular results for $n(k)$ in the CDFM (Antonov et al. 1979, 1980) using the symmetrized Fermi-type distribution in (8.57) for ^{4}He and ^{16}O are given in Fig. 8.6 and Fig. 8.7 respectively and are compared with exp(S)-method results (Zabolitzky and Ey 1978) and with GCM calclations (Antonov et al. 1986a) without the delta-function approximation (see Section 8.2).

The proton momentum distributions of ^{39}K, ^{40}Ca, and ^{48}Ca have been calculated (Antonov et al. 1983b) using model-independent charge distributions obtained from analyses of electron elastic scattering and muonic atoms (Malaguti et al. 1979a) and also using charge distributions calculated from the SPP method (Malaguti et al. 1979a). As an example Fig. 8.9 shows the proton momentum distribution in ^{40}Ca.

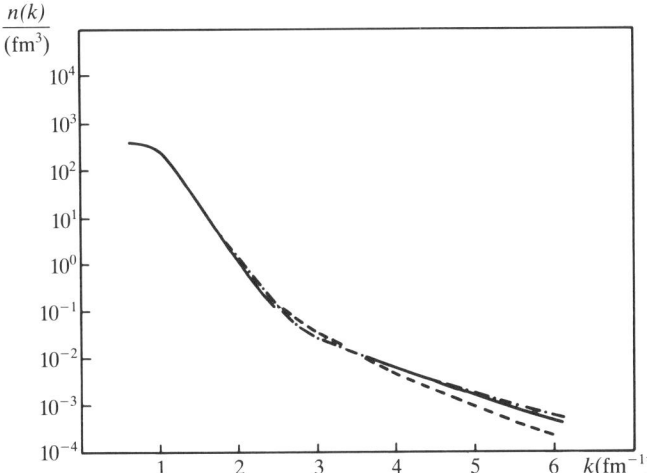

FIG. 8.9. Proton momentum distribution of ^{40}Ca (Antonov et al. 1985). The solid line represents the calculations using SPP charge distribution. The dashed line (Sick I) and dash-dotted line (Sick II) represent calculations using model-independent data for $\rho_{ch}(r)$ taken from the work of Malaguti et al. (1979a).

An extension of the CDFM relations to more general types of density distributions $\rho(r)$ is made in a simple way (Antonov et al. 1986b; Antonov and Petkov 1987). The generalization is carried out for the cases of $\rho(r)$ with one and two maxima. In the case of ^{12}C (Fig. 8.10) the proton momentum distribution has been calculated using the charge density obtained in the SPP method and from experimental data (Reuter et al. 1982). In the same figure the neutron momentum distribution calculated using $\rho_N(r)$ for point neutrons taken from the SPP method is shown.

In Fig. 8.11 the proton momentum distribution for ^{90}Zr is illustrated. The calculations were performed using densities $\rho_{ch}(r)$ taken from the SPP method (Malaguti et al. 1979b) and experimental data taken from the work of Malaguti et al. (1982).

The proton momentum distribution for ^{208}Pb calculated in CDFM (Antonov and Petkov 1987) is shown and compared with other approaches in Fig. 7.12.

It is interesting to note that an inherent feature of the results for $n(k)$ obtained in the CDFM and GCM methods with an infinite square-well construction potential (Antonov et al. 1986a), see Section 8.2, is the presence of high-momentum components at $k > 2\,\text{fm}^{-1}$. This is evidence for the effectiveness of both methods in accounting for short-range

138   OTHER PHENOMENOLOGICAL MODELS

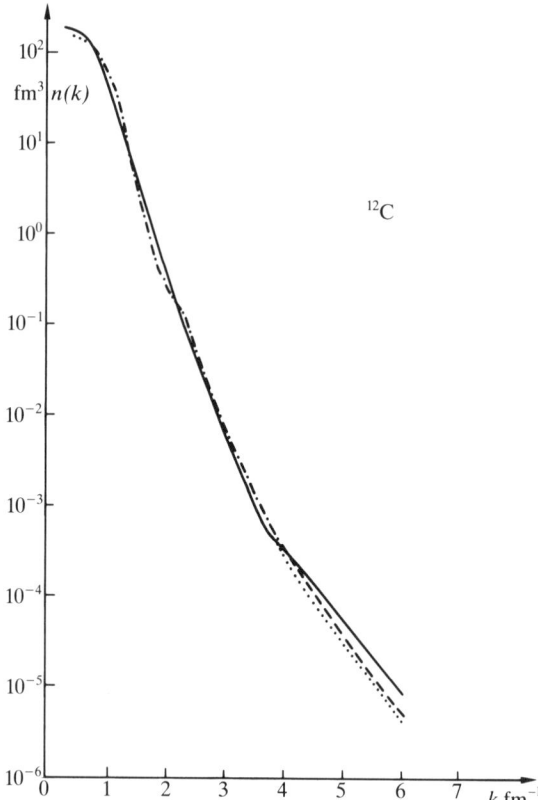

FIG. 8.10. Momentum distribution in ^{12}C calculated in the CDFM (Antonov *et al.* 1986*b*). Proton momentum distribution: using SPP charge distribution (solid line); using $\rho_{ch}(r)$ from experimental data (Reuter *et al.* 1982) (dashed line). Neutron momentum distribution: using SPP point-neutron-density distribution (points).

correlations between nucleons. The common property of these approaches are the intermediate generating states corresponding to configurations (fluctons) in which the nucleons are confined in a finite volume. Some of these configurations have a high density value ($\rho_0(x) \sim 1/x^3$ and small x) so that the particles are close to each other and the short-range components of the nucleon–nucleon forces are operative.

To complete the CDFM we need to determine the weight function $f(x)$ dynamically on the basis of eqn (8.25):

$$-\frac{\hbar^2}{2m_{\text{eff}}}\frac{d^2 f(x)}{dx^2} + V(x)f(x) = Ef(x). \tag{8.62}$$

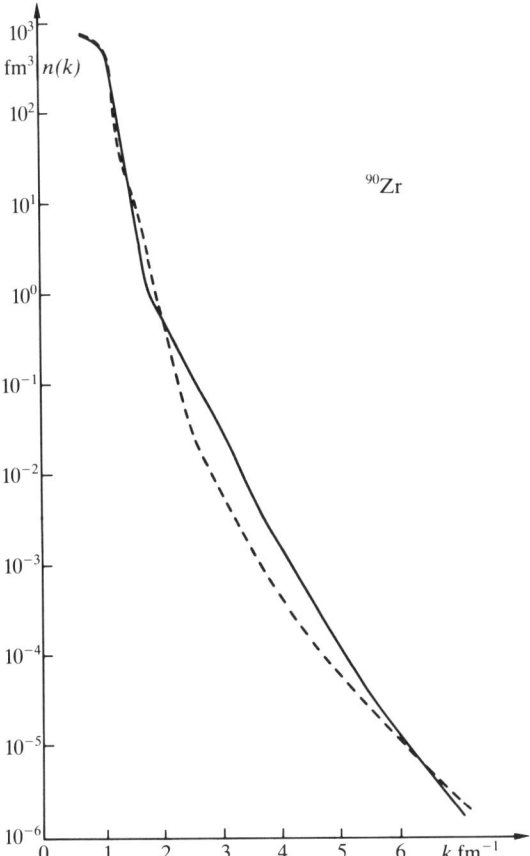

FIG. 8.11. Proton momentum distribution in ^{90}Zr calculated in the CDFM (Antonov *et al.* 1986*b*): using SPP charge distribution (Malaguti *et al.* 1979*b*) (solid line); using experimental data for $\rho_{ch}(r)$ taken from the work of Malaguti *et al.* (1982) (dashed line).

(An analogous equation has been used by G. Brown *et al.* (1983) treating the zero-motion breathing vibrations of the nucleon's three-quark bag).

The function $V(x) = \langle \Phi(\{r_i\}, x) | \hat{H} | \Phi(\{r_i\}, x) \rangle$ corresponds to the potential energy of the coherent-breathing collective motion of all A nucleons. As the intermediate states in the CDFM are uniform sets of A nucleons we use for $V(x)$ the corresponding expression for the energy of nuclear matter with density $\rho_0(x)$ (Brueckner *et al.* 1968):

$$V(x) = AV_0(x) + V_c - V_{ce}, \tag{8.63}$$

where

$$V_0(x) = 37.53[(1+\alpha)^{\frac{5}{3}} + (1-\alpha)^{\frac{5}{3}}]\rho_0^{\frac{2}{3}}(x) + b_1\rho_0(x) + b_2\rho_0^{\frac{4}{3}}(x)$$
$$+ b_3\rho_0^{\frac{5}{3}}(x) + \alpha^2[b_4\rho_0(x) + b_5\rho_0^{\frac{4}{3}}(x) + b_6\rho_0^{\frac{5}{3}}(x)],$$

$$V_c = \frac{3}{5}\frac{Z^2 e^2}{x}, \quad V_{ce} = 0.7386 e^2 Z (3Z/4\pi x^3)^{\frac{1}{3}},$$

$$\alpha = \frac{N-Z}{N+Z}; \quad b_1 = -741.28, \quad b_2 = 1179.89, \quad b_3 = -467.54,$$

$$b_4 = 148.26, \quad b_5 = 372.84, \quad b_6 = -769.57$$

(8.64)

The CDFM equation (8.62) has several solutions corresponding to binding states, whose number depends on the nuclear mass number A. The calculations show that with the particular choice of $V(x)$ (8.63)–(8.64) and at an effective mass $m_{\text{eff}} \cong 3 m_N/A$ the solution of (8.62) reproduces the binding energies of the nuclei and the density distributions $\rho(r)$.

In the essential region near the minimum, the function of the collective potential energy can be approximated fairly well by the Morse potential:

$$V(x) = U_0\{\exp[-2\alpha(x-x_0)] - 2\exp[-\alpha(x-x_0)]\}. \quad (8.65)$$

The solutions of (8.62) for the functions $f_n(x)$ and energies E_n then have well-known analytic forms (Landau and Lifshitz 1977).

The results obtained in the framework of CDFM (Antonov et al. 1985) lead to the following conclusions:

i) the breathing vibrations appearing in the model correspond to a high incompressibility and their energies are comparable with the nuclear binding energies.

ii) The nuclear densities of excited nuclei are generally decreased in the centre relative to those of the ground state, and the influence of the nuclear surface on the collective motion is increased. The nuclear size is essentially larger in the excited states than in the ground state. As an example Fig. 8.12 shows the density distributions for the ground and first excited states of ^{40}Ca.

iii) In the appearance of breathing vibrations CDFM predicts the existence of a threshold with respect to the mass number A. The first excited level appears in all nuclei with $A \geqslant 12$ and the second one appears in the nuclei beyond $A \approx 70$.

A review of CDFM and its applications for calculations of different nuclear ground-state and nuclear process characteristics is given by Antonov et al. (1983a).

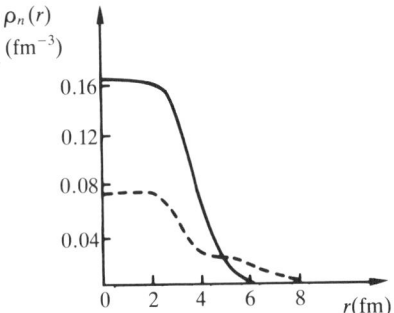

FIG. 8.12. Nucleon density distributions $\rho_0(r)$ (solid line) and $\rho_1(r)$ (dashed line) in ^{40}Ca (Antonov et al. 1985).

8.4. Comparison with experimental data

At present there is no method for directly measuring the nucleon momentum distribution in nuclei. The quantities which are measured by particle–nucleus and nucleus–nucleus collisions are the cross-sections of different reactions, and these must be interpreted using interaction models before we can obtain information about the nucleon momentum distributions.

At projectile energies of order several hundred MeV per particle, the cross-sections contain information on the momentum distributions of the target nucleons. For example, the cross-section for backward emission of particles is determined (Frankel 1977) by the expression:

$$\frac{d\sigma}{d^3q} = \frac{C(p, k_{\min})}{|\mathbf{p} - \mathbf{q}|} \int_{k_{\min}}^{\infty} n(k) k \, dk, \tag{8.66}$$

(see also Section 3.2).

A systematic collection (Jaminon et al. 1986) of the available data from different reactions by means of eqn (8.66) and a comparison with the theoretical predictions of the approach in (Jaminon et al. 1986) (PPW + RPA = dash-dotted line) and CDFM method = dotted line) are shown in Fig. 8.13. Empirical values of $n(k)/Z$ extracted from collisions of high-energy protons with ^{208}Pb (Gurvitz 1982) (short dashes) and of 720 MeV alpha-particles on ^{181}Ta (Avan et al. 1984; Cordell et al. 1981b) (dash-two-dots line), as well as from the reactions (α, ^{3}He), (^{16}O, ^{15}O) (solid line) (Araseki and Fujita 1985; Fujita and Kubodera 1984) and from (π^+, π^+p) and (π^+, 2p) (long dashes) (Grashin and Shalamov 1979) are given in the same figure.

The spread of data for $n(k)$ obtained from different processes shows the difficulties of reliable extraction of $n(k)$ from experimental results.

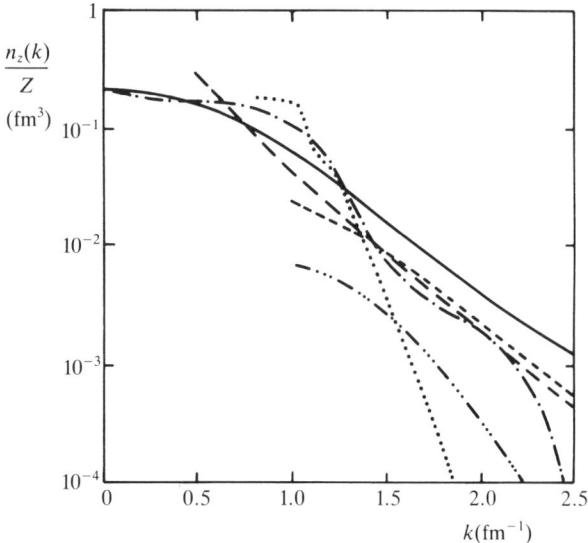

FIG. 8.13. Comparison between the proton momentum distribution in the PPW + RPA model (Jaminon et al. 1986) (dash-dotted line) and the CDFM for ^{208}Pb (Antonov and Petkov 1987) (dotted line) with empirical values extracted from: collisions of high-energy protons with ^{208}Pb (Gurvitz 1982) (short dashes); of alpha-particles on ^{181}Ta (Avan et al. 1984; Cordell et al. 1981b) (dash three dots), as well as from the reactions (α, ^{3}He), (^{16}O, ^{15}O) (Araseki and Fujita 1985; Fujita and Kubodera 1984) (solid line), and (π^+, π^+p), (π^+, $2p$) (Grashin and Shalamov 1979) (long dashes).

Further investigations of different reactions are necessary together with theoretical studies of their mechanism and the calculation of the nucleon momentum distribution itself.

Another class of experiments which gives indirect information on $n(k)$ is the deep-inelastic (e, e'p) reaction. Studies of the (e, e'p) process and the spectral function $S(k, \omega)$ determined experimentally and theoretically are reviewed in the paper of Frullani and Mougey (1984).

As pointed out in Section 1.4 the hole-state spectral function $S_h(k, \omega)$ determines the momentum distribution (1.101):

$$n(k) = \int_{-\infty}^{\varepsilon_F} d\omega' S_h(k, \omega'). \tag{8.67}$$

In the papers of Mougey et al. (1976) and Nakamura et al. (1976) detailed experimental and theoretical analyses of the (e, e'p) reaction on different nuclei have been made. In the work of Mougey et al. (1976) the spectral functions for the nuclei ^{28}Si, ^{40}Ca, and ^{58}Ni have been determined from the (e, e'p) cross-sections using distorted wave-impulse approximation.

A number of theoretical investigations was devoted to the spectral functions of hole nuclear states as a source of information on the momentum distribution, deep-hole energies, and widths etc. (Antonov et al. 1982; Köhler 1966; Gross and Lipperheide 1970; Engelbrecht and Weidenmüller 1972; Sartor 1976, 1977; Sartor and Mahaux 1980b; Boffi et al. 1977; Boffi and Capuzzi 1979, 1981).

Köhler (1966) calculated the 1s single-particle width as a function of nuclear medium density and the momentum of the hole using Brueckner K-matrix theory for infinite systems. Calculations based on the single-particle Green-function formalism (Orland and Schaeffer 1978; Antonov et al. 1982; Sartor and Mahaux 1980b; Boffi et al. 1977; Boffi and Capuzzi 1979, 1981) provide a reasonable explanation of the experimental spectral functions and related quantities.

An explicit expression for the spectral function of the hole nuclear states was obtained in the framework of CDFM (Antonov et al. 1982) in the form:

$$S_h(k, \omega \leq E_F) = \frac{\pi \alpha}{k\sqrt{[\mu(\omega - E_F)]}} \left| f\left[\frac{\alpha}{k}\sqrt{\frac{(\omega - E_F)}{\mu}}\right]\right|^2, \quad (8.68)$$

where f is the weight function of the CDFM, $\alpha = (\frac{19}{8}\pi A)^{\frac{1}{3}}$, the energy E_F is the nucleon-separation energy, the parameter $\mu = -50$ MeV. The values of k are determined from the positions of the maximum of $S_h(k, \omega)$ for each hole nuclear state.

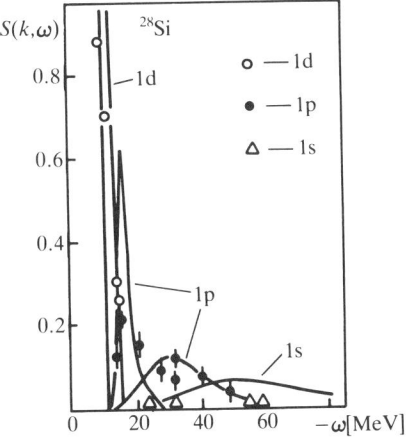

FIG. 8.14. Comparison between empirical spectral functions (Mougey et al. 1976) and that calculated in the CDFM (Antonov et al. 1982) for ^{28}Si. The ordinate scale is in arbitrary units.

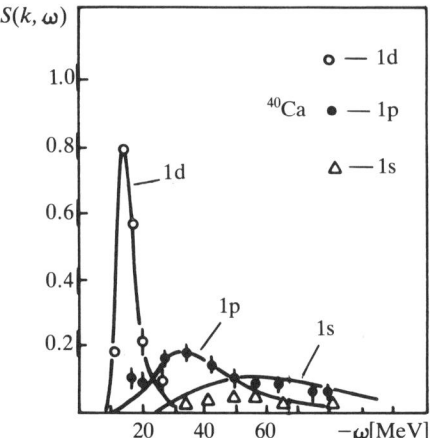

FIG. 8.15. Same as in Fig. 8.14 for ^{40}Ca.

The hole-state spectral functions for ^{28}Si, ^{40}Ca, and ^{58}Ni obtained in CDFM (Antonov *et al.* 1982) are shown in Figs 8.14, 8.15, and 8.16 respectively. It must be noticed that the main experimental regularities of the spectral functions, namely the range of the widths and the asymmetry form of $S_h(k, \omega)$ are described quite well. Note that the spectral function (8.68) is in accord with the momentum distribution $n(k)$ (8.67): they are both expressed by means of the weight function $f(x)$, the latter being related to the local density distribution $\rho(r)$ (8.56)).

The theoretical analysis of the quasi-free (e, e'p) reaction cross-section

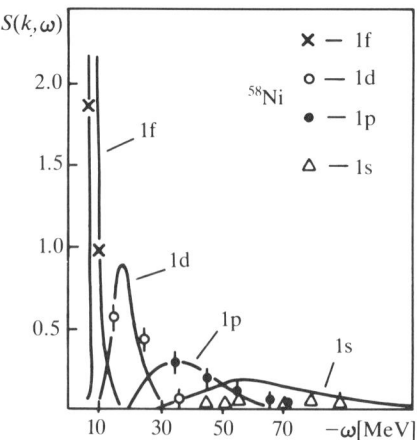

FIG. 8.16. Same as in Fig. 8.14 for ^{58}Ni.

is carried out by Boffi et al. (1982a) in terms of four structure functions. The different behaviour of the structure functions in the reaction $^{16}\text{O}(e, e'p)^{15}\text{N}$ is analyzed for the $p_{\frac{1}{2}}$ and $p_{\frac{3}{2}}$ hole states in ^{16}O in the framework of a generalized distorted-wave impulse approximation.

Another class of experiments, namely (γ, p) and (γ, n) reactions at photon energies above the giant-resonance region provide a tool for studying the high-momentum components of the nuclear wavefunction. The data from these reactions may be used, in principle, to determine to what extent short-range correlations have to be introduced to modify the shell-model description, which is believed to be a good first-order approximation. Reviews on (γ, p) and (γ, n) reactions are given, for instance, in the paper of Frullani and Mougey (1984) and in the book of Barrett and Jackson (1977). The data are usually compared with calculations based on a direct, single-particle knockout mechanism, and on mechanisms involving two particles in an intermediate state. Calculations based on the quasi-free knockout model (QFK) (Boffi et al. 1981) in which the photon interacts with a single nucleon, the other $A - 1$ nucleons being spectators, have suggested that all available (γ, p_0) data for photon energies ~ 120 MeV can be explained in the framework of this model.

An attempt to extract the proton momentum distribution in ^4He from the $^4\text{He}(\gamma, p)^3\text{H}$ reaction (Gorbunov 1968, Kiergan et al. 1973) as well as from $^4\text{He}(e, e'p)^3\text{H}$ experimental data (Goldstein et al. 1981) is given in Fig. 8.17. As can be seen the momentum-distribution behaviour at $p_B > 350$ MeV/c supports the DWIA (distorted-wave impulse approximation) calculations corrected for short-range correlation effects (Goldstein et al. 1981). As shown by Frullani and Mougey (1984) it is possible to determine the momentum distribution of a nucleon in a certain shell from the $(e, e'p)$ and (γ, p) reaction data in the so-called 'extended PWIA'. In Fig. 8.18 the 1p shell momentum distribution for ^{12}C is shown. The $p_{\frac{1}{2}}$ and $p_{\frac{3}{2}}$ shell momentum distributions for ^{16}O deduced from $(e, e'p)$ (Bernheim et al. 1982) and (γ, p) (Findlay and Owens 1977b; Matthews et al. 1977; Leitch 1979; Findlay et al. 1978) reactions are presented in Figs. 8.19 and 8.20 respectively and compared with different theoretical estimations.

The experiments on the $^{16}\text{O}(\gamma, p)^{15}\text{N}$ reaction for $E_\gamma = 100$–400 MeV (Leitch et al. 1985) as well as on $^7\text{Li}(\gamma, p)$ and $^7\text{Li}(e, p)$ reactions (Senè et al. 1985) and $^{40}\text{Ca}(\gamma, p_0)^{39}\text{K}$ at $E_\gamma = 100$–300 MeV (Leitch et al. 1986) extend the range covered by the previous experiments, but the comparison with the available theoretical work still does not allow a definite conclusion about the reaction mechanism. As pointed out in the work of Leitch et al. (1985) none of the calculations carried out in the QFK model (Boffi et al. 1981), in a Jastrow-type model (Weise and Huber 1971), in

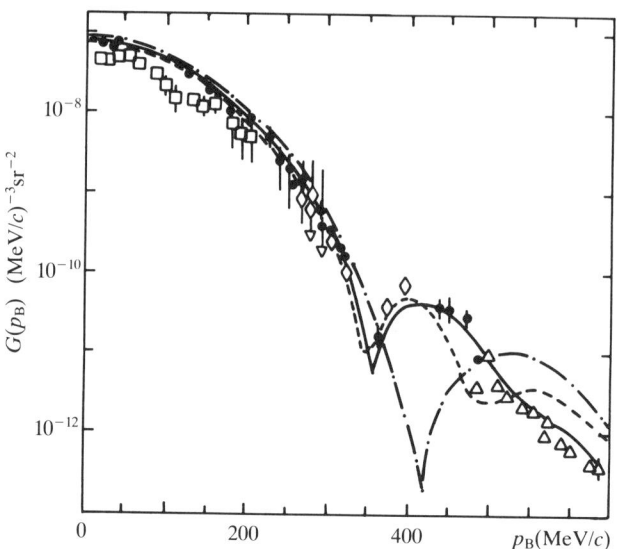

FIG. 8.17. Momentum distributions of protons in ^{4}He deduced from ^{4}He(e, e'p)^{3}H and ^{4}He(γ, p)^{3}H reactions and presented in the review of Frullani and Mougey (1984). (e, e'p) data are obtained by Goldstein et al. (1981) (open squares: kinematics 1; solid circles: kinematics 2). (γ, p) data are obtained by Gorbunov (1968) (open diamonds) and Kiergan et al. (1973) (open triangles). The dash-dotted curve is the PWIA calculation corrected for short-range correlation effects. The dashed curve is the DWIA calculation, while in the solid line correlations are also taken into account (adapted from Goldstein et al. 1981).

an approach with initial- and final-state correlations plus a representation of meson exchange currents (Hebach et al. 1976) and in a self-consistent RPA theory, is able to reproduce all features of the data. The conclusion is drawn by Leitch et al. (1985) that no single mechanism dominates the (γ, N) processes over a large part of the whole energy and angular range, though for example the QFK contribution to these reactions is not negligible. It is shown by Senè et al. (1985) that the quasi-deuteron model (Levinger 1951; Schoch 1978), in which the photon absorption takes place on correlated n-p pairs, or 'quasi-deuterons', is more successful in describing the ^{7}Li(γ, N) data than the simple QFK mechanism, though the contributions from this mechanism are not negligible, particularly at backward angles.

Considerable attention is paid also to the (γ, n) reaction as a potential source of information on nucleon correlations in nuclei (Senè et al. 1985; Göringer et al. 1982; Göringer and Schoch 1980; Schier and Schoch 1974; Tsai et al. 1985; Belyaev et al. 1984). Differential cross-sections

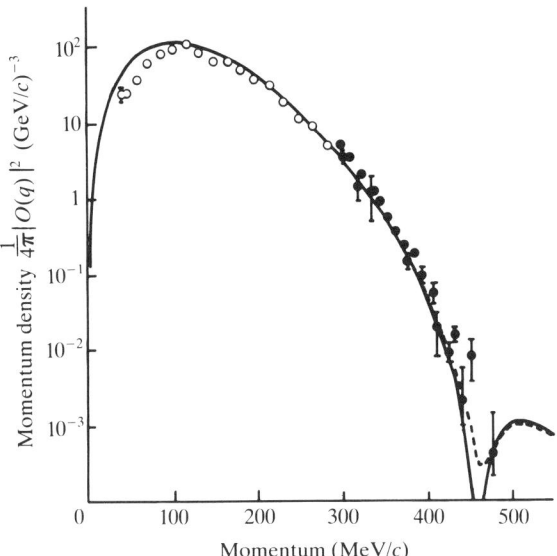

FIG. 8.18. Momentum distribution of 1p-shell nucleons in ^{12}C deduced from (e, e'p) and (γ, p_0) reactions and presented in the review of Frullani and Mougey (1984). (e, e'p) data are obtained by Mougey et al. (1976) (open circles). (γ, p_0) data are obtained by Matthews et al. (1976) (solid circles). The solid line represents the computed $1p_{\frac{3}{2}}$ distribution calculated from the Elton-Swift potential (1967). The dashed line, where distinguishable, shows the effect of a 10% $1p_{\frac{1}{2}}$ admixture (from Findlay and Owens (1977b)).

measured, for instance, for the reactions ^{16}O(γ, n_0)^{15}O (Göringer et al. 1982; Göringer and Schoch 1980), ^{12}C(γ, n)^{11}C (Schier and Schoch 1974), ^{7}Li(γ, n) (Senè et al. 1985), ^{133}Cs(γ, n)^{132}Cs (Tsai et al. 1985), ^{206,208}Pb(γ, n), ^{181}Ta, ^{209}Bi(γ, n) (Belyaev et al. 1984) provide additional information about the high-momentum components of the ground-state nuclear wavefunction.

If the single-particle absorption mechanism in which the cross-section is related to the single-particle proton wavefunction $\Phi(p)$ (Göringer et al. 1982)

$$\frac{d\sigma}{d\Omega}(E_\gamma, \theta) = C(E_\gamma, \theta) |\Phi(p)|^2 \qquad (8.69)$$

dominates, the cross-section for (γ, n) processes have to be much smaller than for the corresponding (γ, p) processes, since in the neutron case the photon interacts only with the magnetic moment or with an 'effective' charge due to the c.m. motion of the neutron. This, however, is not in

OTHER PHENOMENOLOGICAL MODELS

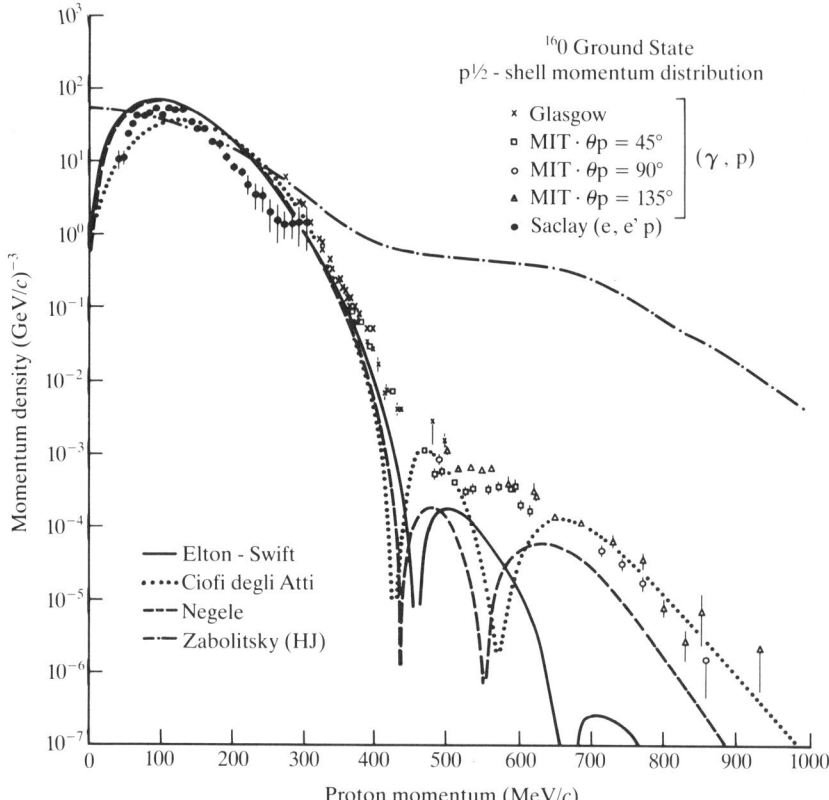

FIG. 8.19. Momentum distribution of $p_{\frac{1}{2}}$-shell protons on ^{16}O deduced from the data of (e, e'p) and (γ, p_0) reactions and presented in the review of Frullani and Mougey (1984). (e, e'p) data are obtained by Bernheim et al. (1982). (γ, p_0) data are taken from the works of Findlay and Owens (1977b), Matthews et al. (1977), and Leitch (1979). The experimental points deduced from the data through an extended PWIA analysis are compared with several theoretical results: i) calculations with the Elton-Swift potential (1967) (solid line); ii) density-dependent Hartree–Fock calculations of Negele (1970) (dashed line); iii) calculations of Ciofi Degli Atti (1971) using harmonic-oscillator wavefunctions with the Jastrow correlation function (dotted line); iv) the dot-dashed line is the result of the calculations of Zabolitzky and Ey (1978) for the entire ^{16}O nucleus in the $\exp(S)$ method using the Hamada–Johnston potential.

agreement with the experimental (γ, n) data (Göringer et al. 1982; Schier and Schoch 1974; Miller et al. 1971; Schier and Schoch 1975). The (γ, n) experiments, although much poorer qualitatively and quantitatively than the corresponding (γ, p) data (Matthews et al. 1976, 1977; Leitch 1979; Findlay and Owens 1977a), already show that (γ, n) and (γ, p) cross-

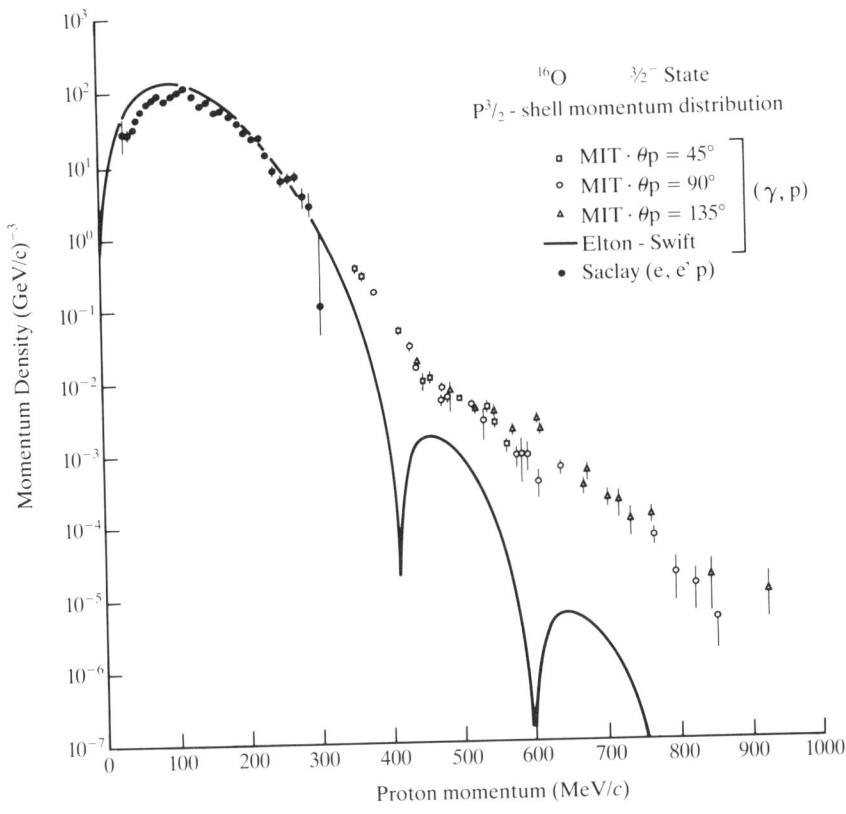

FIG. 8.20. Momentum distribution of $p_{3/2}$-shell protons in ^{16}O deduced from Saclay (e, e'p) data and (γ, p) reactions and presented in the review of Frullani and Mougey (1984). Experimental points from the (γ, p) reaction are deduced by Leitch (1979). The solid line shows the calculations using Elton–Swift wavefunctions (1967).

sections are comparable in size with a similar energy and angle dependence. This leads us to the assumption that two nucleons are involved in the photo-absorption. The (γ, n) data now available can be described qualitatively in the framework of the quasi-deuteron model (Levinger 1951), by the two-step mechanism (Cotanch 1978) consisting of a direct knockout (γ, p) process followed by a (p, n) charge exchange, as well as by the model proposed by Göringer et al. (1982) in which the photo-absorption takes place by a neutron-proton pair with one nucleon leaving the nucleus whereas the other nucleon takes over the momentum difference and remains within the nucleus.

It can be concluded that up to now neither experimental data nor theoretical considerations of the nuclear photo-absorption are sufficiently comprehensive to give reliable information about the momentum distribution of nucleons.

As shown by Amado and Woloshyn (1977a), the inclusion of the final-state interaction leads to difficulties in the extraction of information on $n(k)$ from high-energy experiments (see also Section 3.2). This problem has been thoroughly investigated from a theoretical point of view (Amado 1977; Noble 1978; Eisenberg *et al.* 1979). In the case of $(e, e'p)$ and (γ, p) reactions the problem of the lack of orthogonality between continuum final states and the initial bound states has been discussed by Boffi *et al.* (1982*b*) who show that the spurious effect originating from an orthogonality defect is particularly important when the momentum transfer is small compared with the momentum of the outgoing proton. As pointed out by Boffi *et al.* (1982*b*) the problem is practically absent in the case of $(e, e'p)$ reactions in typical quasi-free kinematical conditions, while in the (γ, p) reaction at photon energies of about 100 MeV the lack of orthogonality can introduce errors when the reaction is analysed as a one-step process.

REFERENCES

Abrikosov, A. A., Gorkov, L. P., and Dzyaloshinsky, I. E. (1963). *Methods of quantum field theory in statistical physics.* Prentice-Hall, Englewood Cliffs.
Amado, R. D. (1976). *Physical Review,* **C14,** 1264.
Amado, R. D. (1977). *Physical Review,* **A16,** 1725.
Amado, R. D. and Woloshyn, R. M. (1976a). *Physical Review Letters,* **36,** 1435.
Amado, R. D. and Woloshyn, R. M. (1976b). *Physics Letters,* **62B,** 253.
Amado, R. D. and Woloshyn, R. M. (1977a). *Physics Letters,* **69B,** 400.
Amado, R. D. and Woloshyn, R. M. (1977b). *Physical Review,* **C15,** 2200.
Antonov, A. N. and Petkov, I. Zh. (1986). *Nuovo Cimento,* **94A,** 68.
Antonov, A. N. and Petkov, I. Zh. (1987). *Bulgarian Journal of Physics,* **14,** 137.
Antonov, A. N., Christov, Chr. V., and Petkov, I. Zh. (1986a). *Nuovo Cimento,* **91A,** 119.
Antonov, A. N., Hodgson, P. E., and Petkov, I. Zh. (1987) *Nuovo Cimento,* **A** (in press).
Antonov, A. N., Nikolaev, V. A., and Petkov, I. Zh. (1979). *Bulgarian Journal of Physics,* **6,** 151.
Antonov, A. N., Nikolaev, V. A., and Petkov, I. Zh. (1980). *Zeitschrift für Physik,* **A297,** 257.
Antonov, A. N., Nikolaev, V. A., and Petkov, I. Zh. (1982). *Zeitschrift für Physik,* **A304,** 239.
Antonov, A. N., Nikolaev, V. A., and Petkov, I. Zh. (1983a). *Izv. AN USSR, ser. fiz.* **47,** 134.
Antonov, A. N., Nikolaev, V. A., and Petkov, I. Zh. (1985). *Nuovo Cimento,* **86A,** 23.
Antonov, A. N., Petkov, I. Zh., and Hodgson, P. E. (1986b). *Bulgarian Journal of Physics,* **13,** 110.
Antonov, A. N., Nikolaev, V. A., Petkov, I. Zh., and Hodgson, P. E. (1983b). *Bulgarian Journal of Physics,* **10,** 590.
Araseki, H. and Fujita, T. (1985). *Nuclear Physics,* **A439,** 681.
Avan, M., Baldit, A., Castor, J., Chaigne, G., Devaux, A., Fargeix, J., Force, P., Landdaud, C., Roche, C., Vicente, J., Didelez, J. P., Reide, F., and Gurvitz, S. A. (1984). *Physical Review,* **C30,** 521.
Azhgirey, L. S., Vzorov, I. K., Zhmyrov, V. N., Ivanov, V. V., Ignatenko, M. A., Kuznetzov, A. S., Mescheryakov, M. G., Razin, S. V., and Stoletov, G. D. (1977). *JINR,* I-10842, Dubna.
Baranger, M. (1970). *Nuclear Physics,* **A149,** 225.
Barrett, R. C. and Jackson, D. F. (1977). *Nuclear sizes and structure.* Clarendon, Oxford.
Bartel, J., Brack, M., and Durand, M. (1985). *Nuclear Physics,* **A445,** 263.

Bartz, B. I., Bolotin, Yu. L., Inopin, E. V., and Gonchar, V. Yu. (1982). *The Hartree–Fock method in the nuclear theory*. Naukova Dumka, Kiev.
Beiner, M., Flocard, H., Van Giai, N., and Quentin, P. (1975). *Nuclear Physics*, **A238**, 29.
Belyaev, S. N., Vasilyev, O. V., Kozin, A. B., Nechkin, A. A., and Semyonov, V. A. (1984). *Izv. AN USSR, ser. fiz.* **48**, 1940.
Belyakov, V. A. (1961). *ZETP USSR*, **40**, 1210; *Soviet Physics, JETP*, **15**, 850.
Benhar, O., Ciofi Degli Atti, C., Fantoni, S., and Rosati, S. (1979). *Nuclear Physics*, **A328**, 127.
Benhar, O., Ciofi Degli Atti, C., Liuti, S., and Salmè, G. (1986). *Physics Letters*, **117B**, 135.
Bernheim, M., Bussiere, A., Mougey, J., Royer, D., Tarnowski, D., Turck-Chieze, S., Frullani, S., Capitani, G. P., de Sanctis, E., and Jans, E. (1981). *Nuclear Physics*, **A365**, 349.
Bernheim, M., Bussiere, A., Mougey, J., Royer, D., Tarnowski, D., Turck-Chieze, S., Frullani, S., Boffi, S., Giusti, C., Pacati, F. D., Capitani, G. P., de Sanctis, E., and Wagner, G. J. (1982). *Nuclear Physics*, **A375**, 381.
Bertch, G. F. (1981). *Physical Review Letters*, **46**, 472.
Bertozzi, W., Friar, J., Heisenberg, J., and Negele, J. W. (1972). *Physics Letters*, **41B**, 408.
Bethe, H. (1968). *Physics Review*, **167**, 879.
Bethe, H. (1971). *Annual Review of Nuclear Science*, **21**, 93.
Bethe, H., Brandow, B. H., and Betschek, A. G. (1963). *Physical Review*, **129**, 225.
Blin, A. H., Hiller, B., Hasse, R. W., and Schuck, P. (1984). *Journal de Physics*, **45**, C6–231.
Boal, D. H. (1980). *Physical Review*, **C21**, 1913.
Boal, D. H. and Woloshyn, R. M. (1979). *Physical Review*, **C20**, 1878.
Bodek, A. and Ritchie, J. L. (1981). *Physical Review*, **D23**, 1070.
Boffi, S. and Capuzzi, F. (1979). *Lett. Nuovo Cimento*, **25**, 209.
Boffi, S. and Capuzzi, F. (1981). *Nuclear Physics*, **A351**, 219.
Boffi, S., Giusti, C., and Pacati, F. D. (1977). *Lettere Nuovo Cimento*, **19**, 594.
Boffi, S., Giusti, C., and Pacati, F. D. (1981). *Nuclear Physics*, **A359**, 81.
Boffi, S., Giusti, C., and Pacati, F. D. (1982a). *Nuclear Physics*, **A386**, 599.
Boffi, S., Cannata, F., Capuzzi, F., Giusti, C., and Pacati, F. D. (1982b). *Nuclear Physics*, **A379**, 509.
Bohigas, O. and Stringari, S. (1980). *Physics Letters*, **95B**, 9.
Bohigas, O., Campi, X., Krivine, H., and Treiner, J. (1976). *Physics Letters*, **64B**, 381.
Brack, M., Guet, C., and Håkansson, H.-B. (1985). *Physics Reports*, **123**, 275.
Brink, D. M. (1966). *Proceedings of the School of Physics, 'Enrico Fermi', course 36* (ed. C. Bloch), Academic Press, New York.
Brink, D. M. and Grypeos, M. E. (1967). *Nuclear Physics*, **A97**, 81.
Brown, B. A., Massen, S. E., Escudero, J. I., Hodgson, P. E., Madurga, G., and Viñas, J. (1983). *Journal of Physics*, **G9**, 423.
Brown, G. E., Durso, J. W., and Johnson, M. B. (1983). *Nuclear Physics*, **A397**, 447.
Brueckner, K. A., Eden, R. J., and Francis, N. C. (1955). *Physical Review*, **98**, 1445.

REFERENCES

Brueckner, K. A., Lin, W.-F., and Lombard, R. J. (1969a). *Physical Review*, **181**, 1506.
Brueckner, K. A., Lin, W.-F., Lombard, R. J., and Clark, R. C. (1970). *Physical Review*, **C1**, 249.
Brueckner, K. A., Buchler, J. R., Jorna, S., and Lombard, R. J. (1968). *Physical Review*, **171**, 1188.
Brueckner, K. A., Buchler, J. R., Clark, R. C., and Lombard, R. J. (1969b). *Physical Review*, **181**, 1543.
Burov, V. V., Lukyanov, V. K., and Titov, A. I. (1977). *Physics Letters*, **67B**, 46.
Burov, V. V., Eldyshev, Yu. N., Lukyanov, V. K., and Pol', Yu. S. (1974). *JINR*, E4-8029, Dubna.
Calogero, F. and Degasperis, A. (1975). *Physical Review*, **A11**, 265.
Campi, X. and Sprung, D. W. (1972). *Nuclear Physics*, **A194**, 401.
Campi, X., Sprung, D. W., and Martorell, J. (1974). *Nuclear Physics*, **A223**, 541.
Casas, M., Martorell, J., and Moya de Guerra, E. (1986). *Physics Letters*, **167B**, 263.
Cavinato, M., Marangoni, M., and Sarius, A. M. (1983). *Nuovo Cimento*, **76A**, 197.
Cavinato, M., Marangoni, M., and Sarius, A. M. (1984a). *Nuclear Physics*, **A422**, 237.
Cavinato, M., Marangoni, M., and Sarius, A. M. (1984b). *Nuclear Physics*, **A422**, 273.
Cavinato, M., Marangoni, M., Ottavini, P. L., and Sarius, A. M. (1982). *Nuclear Physics*, **A373**, 445.
Chandra, H. and Sauer, G. (1976). *Physical Review*, **C13**, 245.
Ciofi Degli Atti, C. (1971). *Lett. Nuovo Cimento*, **1**, 590.
Ciofi Degli Atti, C., Pace, E., and Salmè, G. (1984). *Physics Letters*, **141B**, 14.
Clark, J. W. (1981). In *The many-body problem. Jastorw correlations versus Brueckner theory*, (eds. R. Guardiola and J. Ros), p. 184. Springer-Verlag, Berlin.
Clement, C. F. (1969). *Physics Letters*, **28B**, 395, 398.
Clement, C. F. (1973). *Nuclear Physics*, **A213**, 493, 469, 510.
Cordell, K. R., Thornton, S. T., Dennis, L. C., Doering, R. R., Parks, R. L., and Schweizer, T. C. (1981a). *Nuclear Physics*, **A352**, 485.
Cordell, K. R., Thornton, S. T., Dennis, L. C., Doering, R. R., Parks, R. L., and Schweizer, T. C. (1981b). *Nuclear Physics*, **A362**, 431.
Cotanch, S. R. (1978). *Physics Letters*, **76B**, 19.
Czyż, W. and Gottfried, K. (1961). *Nuclear Physics*, **21**, 676.
Da Providencia, J. and Shakin, C. M. (1964). *Annals of Physics*, **30**, 95.
Da Providencia, J. and Shakin, C. M. (1965). *Nuclear Physics*, **65**, 54.
Dal Rì, M., Stringari, S., and Bohigas, O. (1982). *Nuclear Physics*, **A376**, 81.
De Forest, T. and Walecka, J. D. (1966). *Advances in Physics*, **15**, 1.
Dechargè, J. and Gogny, D. (1980). *Physical Review*, **C21**, 1568.
Dechargè, J. and Sips, L. (1983). *Nuclear Physics*, **A407**, 1.
Dechargè, J., Girod, M., and Gogny, D. (1978). Proceedings of the Conference on Modern Trends in Elastic Electron Scattering, Amsterdam.
Dellagiacoma, F., Orlandini, G., and Traini, M. (1983). *Nuclear Physics*. **A393**, 95.

Dieperink, A. E. L., De Forest, T., Sick, I., and Brandenburg, R. A. (1976). *Physics Letters*, **63B,** 261.
Dirac, P. A. M. (1930). *Proceedings of Cambridge Philosophical Society*, **26,** 376.
Eisenberg, J. M., Noble, J. V., and Weber, H. J. (1979). *Physical Review*, **C19,** 276.
Elliot, J. P. and Skyrme, T. H. R. (1953). *Proceedings of the Royal Society of London* ser. A, **232,** 561.
Elton, L. R. B. and Swift, A. (1967). *Nuclear Physics*, **A94,** 52.
Elton, L. R. B. and Webb, S. J. (1970). *Physical Review Letters*, **24,** 145.
Emrich, K., Lührmann, K. H., and Zabolitzky, J. G. (1977). *Physical Review*, **C16,** 1650.
Engelbrecht, C. A. and Weidenmüller, H. A. (1972). *Nuclear Physics*, **A184,** 385.
Fantoni, S. (1978). *Nuovo Cimento*, **A44,** 191.
Fantoni, S. and Pandharipande, V. R. (1984). *Nuclear Physics*, **A427,** 473.
Fantoni, S. and Rosati, S. (1974). *Nuovo Cimento*, **A20,** 179.
Fantoni, S. and Rosati, S. (1975). *Nuovo Cimento*, **A25,** 593.
Fermi, E. (1928). *Zeitschrift für Physik*, **48,** 73.
Findlay, D. J. S. and Owens, R. O. (1977a). *Nuclear Physics*, **A279,** 385.
Findlay, D. J. S. and Owens, R. O. (1977b). *Nuclear Physics*, **A292,** 53.
Findlay, D. J. S., Owens, R. O., Leitch, M. J., Matthews, J. L., Peridier, C. A., Roberts, B. L., and Sargent, C. P. (1978). *Physics Letters*, **74B,** 305.
Flocard, H. and Vautherin, D. (1975). *Physics Letters*, **55B,** 259.
Flocard, H. and Vautherin, D. (1976). *Nuclear Physics*, **A264,** 197.
Flynn, M. F., Clark, J. W., Panoff, R. M., Bohigas, O., and Stringari, S. (1984). *Nuclear Physics*, **A427,** 253.
Frankel, S. (1977). *Physical Review Letters*, **38,** 1338.
Frankel, S. (1978). *Physical Review*, **C17,** 691.
Frankel, S., Frati, W., Woloshyn, R. M., and Yang, D. (1978a). *Physical Review*, **C18,** 1379.
Frankel, S., Frati, W., Van Dyck, O., Werbeck, R., and Highland, V. (1976). *Physical Review Letters*, **36,** 642.
Frankel, S., Frati, W., Blanpied, C., Hoffmann, G. W., Kozlowski, T., Morris, C., Thiessen, H. A., Van Dyck, O., Ridge, R., and Whitten, C. (1978b). *Physical Review*, **C18,** 1375.
French, J. B. and MacFarlane, M. H. (1961). *Nuclear Physics*, **26,** 168.
Friar, J. L. and Negele, J. W. (1973). *Nuclear Physics*, **A212,** 93.
Friar, J. L. and Negele, J. W. (1975). *Nuclear Physics*, **A240,** 301.
Frois, B. (1979). In *Nuclear physics with electromagnetic interactions*. Lectures Notes in Physics Vol. 108 (ed H. Arenhovel and D. Drechsel), p. 52, Springer-Verlag, Berlin.
Frois, B., Bellicard, J. B., Cavedon, J. M., Huet, M., Leconte, P., Ludeau, P., Nakada, A., Ho, P. Z., and Sick, I. (1977). *Physical Review Letters*, **38,** 152.
Frois, B., Cavedon, J. M., Goutte, D., Huet, M., Leconte, P., Papanicolas, C. N., Phan, X. H., Platchkov, S. K., and Williamson, S. E. (1983). *Nuclear Physics*, **A396,** 409c.
Frullani, S. and Mougey, J. (1984). *Advances in Nuclear Physics*, **14,** 1.
Fujita, T. (1977). *Physical Review Letters*, **39,** 174.
Fujita, T. (1986). *Nuclear Physics*, **A457,** 657.

REFERENCES

Fujita, T. and Hüfner, J. (1980). *Nuclear Physics,* **A343,** 493.
Fujita, T. and Kubodera, K. (1984). *Physics Letters,* **149B,** 451.
Galitski, V. M. (1958). *Soviet Physics JETP,* **7,** 104.
Galitski, V. M. and Migdal, A. B. (1958). *Soviet Physics JETP,* **7,** 96.
Gamba, S., Ricco, G., and Rottigni, G. (1973). *Nuclear Physics,* **A213,** 383.
Gari, M., Hyuga, H., and Zabolitzky, J. G. (1976). *Nuclear Physics,* **A271,** 365.
Gaudin, M., Gillespie, J., and Ripka, G. (1971). *Nuclear Physics,* **A176,** 237.
Ghosh, S. K. and Deb, B. M. (1982). *Physics Reports,* **92,** 1.
Ghosh, S. K., Hasse, R. W., Schuck, P., and Winter, J. (1983). *Physical Review Letters,* **50,** 1250.
Gogny, D. (1979). In *Nuclear Physics with electromagnetic interaction.* Lectures Notes in Physics Vol. 108 (ed H. Arenhovel and D. Drechsel), p. 88, Springer-Verlag, Berlin.
Gogny, D. and Padjen, R. (1977). *Nuclear Physics,* **A293,** 365.
Goldstein, V. A., Kuplennikov, E. L., Jibuti, R. I., and Keserashvili, R. Ya. (1981). *Nuclear Physics,* **A355,** 333.
Gorbunov, A. N. (1968). *Physics Letters,* **27B,** 436.
Göringer, H. and Schoch, B. (1980). *Physics Letters,* **97B,** 41.
Göringer, H., Schoch, B., and Lührs, G. (1982). *Nuclear Physics,* **A384,** 414.
Gottfried, K. (1963). *Annals of Physics,* **21,** 29.
Grashin, A. F. and Shalamov, Y. Y. (1979). *Soviet Journal of Nuclear Physics,* **29,** 322.
Griffin, J. J. and Wheeler, J. A. (1957). *Physics Review,* **108,** 311.
Gross, D. H. E. and Lipperheide, R. (1970). *Nuclear Physics,* **A150,** 449.
Guet, C. and Brack, M. (1980). *Zeitschrift für Physik,* **A297,** 247.
Gupta, U. and Rajagopal, A. K. (1982). *Physics Reports,* **87,** 259.
Gurvitz, S. A. (1982). Weizmann Institute of Science Report, WIS-82/7.
Hasse, R. and Schuck, P. (1985*a*). *Nuclear Physics,* **A438,** 157.
Hasse, R. and Schuck, P. (1985*b*). *Nuclear Physics,* **A445,** 205.
Hatch, R. L. and Koonin, S. E. (1979). *Physics Letters,* **81B,** 1.
Hebach, H., Wortberg, A., and Gari, M. (1976). *Nuclear Physics,* **A267,** 425.
Hill, D. L. and Wheeler, J. A. (1953). *Physical Review,* **89,** 1102.
Hiller, B. and Hüfner, J. (1982). *Nuclear Physics,* **A382,** 542.
Hohenberg, P. and Kohn, W. (1964). *Physical Review,* **136,** B864.
Hüfner, J. and Nemes, M. C. (1981). *Physical Review,* **C23,** 2538.
Hugenholtz, N. M. (1957). *Physica (Utrecht),* **23,** 481.
Jaminon, M., Mahaux, C., and Ngô, H. (1985*a*). *Physics Letters,* **158B,** 103.
Jaminon, M., Mahaux, C., and Ngô, H. (1985*b*). *Nuclear Physics,* **A440,** 228.
Jaminon, M., Mahaux, C., and Ngô, H. (1986). *Nuclear Physics,* **A452,** 445.
Jastrow, R. (1955). *Physical Review,* **98,** 1479.
Jeukenne, J. P., Lejeune, A., and Mahaux, C. (1976). *Physics Reports,* **25C,** 83.
Khodel, V. A. and Saperstein, E. E. (1982). *Physics Reports,* **92,** 183.
Kiergan, S. E., Hansen, A. O., and Koester, L. J., Jr. (1973). *Physical Review,* **C8,** 431.
Kirzhnitz, D. A. (1967). *Field theoretical methods in many-body systems.* Pergamon, Oxford.
Kobe, D. H. (1969). *Journal of Chemical Physics,* **50,** 5183.
Köhler, H. S. (1966). *Nuclear Physics,* **88,** 529.
Köhler, H. S. (1975). *Physics Reports,* **18C,** 217.

Kohn, W. and Sham, L. J. (1965). *Physical Review*, **140**, A1133.
Komarov, V. I., Kosarev, G. E., Müller, H., Netzband, D., Stiehler, T. and Tesch, S. (1978). *JINR*, E1-11354, Dubna.
Komarov, V. I., Kosarev, G. E., Müller, H., Netzband, D., Stiehler, T., and Tesch, S. (1979). *JINR*, E1-12393, Dubna.
Krivine, H. (1986). *Nuclear Physics*, **A457**, 125.
Krivine, H., Treiner, J., and Bohigas, O. (1980). *Nuclear Physics*, **A336**, 155.
Kümmel, H. (1971). *Nuclear Physics*, **A176**, 205.
Kümmel, H. and Lührmann, K. H. (1972). *Nuclear Physics*, **A191**, 525.
Kümmel, H. and Zabolitzky, J. G. (1973). *Physical Review*, **C7**, 547.
Kümmel, H., Lührmann, K. H., and Zabolitzky, J. G. (1978). *Physics Reports*, **36**, 1.
Kuo, T. T. S. and Brown, G. E. (1966). *Nuclear Physics*, **85**, 40.
Kuo, T. T. S., Lee, S. Y., and Ratcliff, K. F. (1971). *Nuclear Physics*, **A176**, 65.
Kutzelnigg, W. and Smith, V. H., Jr. (1964). *Journal of Chemical Physics*, **41**, 896.
Lagaris, I. E. and Pandharipande, V. R. (1981). *Nuclear Physics*, **A359**, 349.
Landau, L. D. and Lifshitz, E. M. (1977). *Quantum Mechanics* (3rd edn). Pergamon, Oxford.
Landau, R. H. (1978). *Physical Review*, **C17**, 2144.
Leitch, M. J. (1979). Unpublished Ph.D. thesis. Massachusetts Institute of Technology.
Leitch, M. J., Lin, F. C., Matthews, J. L., Sapp, W. W., Sargent, C. P., Findlay, D. J. S., Owens, R. O., and Roberts, B. L. (1986). *Physics Review*, **C33**, 1511.
Leitch, M. J., Matthews, J. L., Sapp, W. W., Sargent, C. P., Wood, S. A., Findlay, D. J. S., Owens, R. O., and Roberts, B. L. (1985). *Physical Review*, **C31**, 1633.
Levinger, J. S. (1951). *Physical Review*, **84**, 43.
Lifshitz, M. and Singer, P. (1978). *Physical Review*, **41**, 18.
Lipkin, H. J. (1958). *Physical Review*, **110**, 1395.
Lombard, R. J. (1973). *Annals of Physics*, **77**, 380.
Löwdin, P.-O. (1955). *Physical Review*, **97**, 1474.
Lührmann, K. H. and Kümmel, H. (1973). *Nuclear Physics*, **A194**, 225.
Luttinger, J. M. (1961). *Physical Review*, **121**, 942.
Mahaux, C., Bortignon, P. F., Broglia, R. A., and Dasso, C. H. (1985). *Physics Reports*, **120**, 1.
Malaguti, F. (1978). *Nuclear Physics*, **A308**, 125.
Malaguti, F. and Hodgson, P. E. (1973). *Nuclear Physics*, **A215**, 243.
Malaguti, F., Uguzzoni, A., Verondini, E., and Hodgson, P. E. (1978). *Nuclear Physics*, **A297**, 287.
Malaguti, F., Uguzzoni, A., Verondini, E., and Hodgson, P. E. (1979a). *Nuovo Cimento*, **49A**, 412.
Malaguti, F., Uguzzoni, A., Verondini, E., and Hodgson, P. E. (1979b). *Nuovo Cimento*, **53A**, 1.
Malaguti, F., Uguzzoni, A., Verondini, E., and Hodgson, P. E. (1982). *Rivista Nuovo Cimento*, **5**, ser. 3, No. 1.
Małecki, A. and Picchi, P. (1970). *Rivista Nuovo Cimento*, **2**, 119.
Małecki, A. and Picchi, P. (1971). *Physics Letters*, **36B**, 61.
Małecki, A. and Picchi, P. (1973). *Lettere Nuovo Cimento*, **8**, 16.
March, N. H. (1979). *Journal of Chemical Physics*, **71**, 1004.

REFERENCES

Matthews, J. L., Findlay, D. J. S., Gardiner, S. N., and Owens, R. O. (1976). *Nuclear Physics,* **A267,** 51.
Matthews, J. L., Bertozzi, W., Leitch, M. J., Peridier, C. A., Roberts, B. L., Sargent, C. P., Turchinetz, W., Findlay, D. J. S., and Owens, R. O. (1977). *Physical Review Letters,* **38,** 8.
McGuire, J. B. (1964). *Journal of Mathematical Physics,* **5,** 622.
Migdal, A. B. (1957). *Soviet Physics JETP,* **5,** 333.
Migdal, A. B. (1962). *Nuclear Physics,* **30,** 239.
Migdal, A. B. (1967). *Theory of finite fermi-systems and applications to atomic nuclei.* Wiley Interscience, New York.
Millener, D. J. and Hodgson, P. E. (1973). *Nuclear Physics,* **A209,** 59.
Miller, H. G., Buss, W., and Rawlins, J. A. (1971). *Nuclear Physics,* **A163,** 637.
Moszkowski, S. and Scott, B. (1960). *Annals of Physics,* **11,** 65.
Mougey, J., Bernheim, M., De Nercy A. Bussiere, Gillebert, A., Ho Xuan Phan, Priou, M., Royer, D., Sick, I., and Wagner, G. J. (1976). *Nuclear Physics,* **A262,** 461.
Nagorny, S. I., Kasatkin, Yu. A., Kirichenko, I. K., and Inopin, E. V. (1985). *Journal of Nuclear Physics (USSR),* **42,** 870.
Nakamura, K., Hiramatsu, S., Kamae, T., Miramatsu, H., Izutsu, M., and Watase, Y. (1976). *Nuclear Physics,* **A271,** 221.
Narodetsky, I. M. and Simonov, Yu. A. (1975). *Physics Letters,* **58B,** 125.
Negele, J. W. (1970). *Physical Review,* **C1,** 1260.
Noble, J. V. (1978). *Physical Review,* **C17,** 2151.
Orland, H. and Schaeffer, R. (1978). *Nuclear Physics,* **A299,** 442.
Pandharipande, V. R., Papanicolas, C. N., and Wambach, J. (1984). *Physical Review Letters,* **53,** 1133.
Perey, F. (1963a). *Physical Review,* **131,** 745.
Perey, F. (1963b). In *Direct interaction and nuclear reaction mechanism,* p. 125. New York.
Perey, F. and Buck, B. (1962). *Nuclear Physics,* **32,** 353.
Reuter, W., Fricke, G., Merle, K., and Miska, H. (1982). *Physical Review,* **C26,** 806.
Ring, P. and Schuck, P. (1980). *The nuclear many-body problem.* Springer-Verlag, New York.
Ristig, M. L. and Clark, J. W. (1976). *Physical Review,* **B14,** 2875.
Rothhaas, H. (1978). In Proceedings of the Conference on Modern Trends in Elastic Electron Scattering, p. 135. Amsterdam.
Roy, G., Greeniaus, L. G., Moss, G. A., Hutcheon, D. A., Liljestrand, R. L., Woloshyn, R. M., Boal, D. H., Stetz, A. W., Aniol, K., Willis, A., Willis, N., and Camis, R. Mc. (1981). *Physical Review,* **C23,** 1671.
Sartor, R. (1976). *Nuclear Physics,* **A267,** 29.
Sartor, R. (1977). *Nuclear Physics,* **A289,** 329.
Sartor, R. and Mahaux, C. (1980a). *Physical Review,* **C21,** 1546.
Sartor, R. and Mahaux, C. (1980b). *Physical Review,* **C21,** 2613.
Schäfer, L. and Weidenmüller, H. A. (1971). *Nuclear Physics,* **A174,** 1.
Schiavilla, R., Pandharipande, V. R., and Wiringa, R. B. (1986). *Nuclear Physics,* **A449,** 219.
Schier, H. and Schoch, B. (1974). *Nuclear Physics,* **A229,** 93.
Schier, H. and Schoch, B. (1975). *Lettere Nuovo Cimento,* **12,** 334.

Schoch, B. (1978). *Physical Review Letters*, **41**, 809.
Senè, M. R., Anthony, I., Branford, D., Flowers, A. G., Shotter, A. C., Zimmerman, C. H., McGeorge, J. C., Owens, R. O., and Thorley, P. J. (1985). *Nuclear Physics*, **A442**, 215.
Sick, I (1974). *Physics Letters*, **53B**, 15.
Sick, I., Bellicard, J. B., Cavedon, J. M., Frois, B., Huet, M., Leconte, P., Ho, P. X., and Platchkov, S. (1979). Conference on Nuclear Physics with Electromagnetic Interactions.
Siemens, P. J. (1970). *Physical Review*, **C1**, 98.
Sprung, D. W. L. (1972). *Advances in Nuclear Physics*, **5**, 225.
Sprung, D. W. L., Martorell, J., and Campi, X. (1976). *Nuclear Physics*, **A268**, 301.
Svenne, J. P. (1979). *Advances in Nuclear Physics*, **11**, 179.
Thomas, L. H. (1927). *Cambridge Philosophical Society*, **23**, 542.
Thouless, D. J. (1972). *The quantum mechanics of many-body systems*. (2nd edn). Academic Press, New York.
Traini, M. and Orlandini, G. (1985). *Zeitschrift für Physik*, **A321**, 479.
Tsai, J.-S., Prestwich, W. V., and Kennet, T. J. (1985). *Zeitschrift für Physik*, **A322**, 597.
Van Orden, J. W., Truex, W., and Banerjee, M. K. (1980). *Physical Review*, **C21**, 2628.
Vautherin, D. and Brink, D. M. (1972). *Physical Review*, **C5**, 626.
Villars, F. (1963). In Proceedings of the International School of Physics 'Enrico Fermi', course 23. Academic Press, New York.
Weber, H. J. and Miller, L. D. (1977). *Physical Review*, **C16**, 726.
Weise, W. and Huber, M. G. (1971). *Nuclear Physics*, **A162**, 330.
Wildermuth, K. and Tang, Y. C. (1977). *A unified theory of the nucleus*. Vieweg, Braunschweig.
Wong, C. W. (1975). *Physics Reports*, **15**, 283.
Zabolitzky, J. G. (1973). *Physics Letters*, **47B**, 487.
Zabolitzky, J. G. (1974a). *Nuclear Physics*, **A228**, 272.
Zabolitzky, J. G. (1974b). *Nuclear Physics*, **A228**, 285.
Zabolitzky, J. G. and Ey, W. (1978). *Physics Letters*, **76B**, 527.
Zverev, M. V. and Saperstein, E. E. (1986). *Journal of Nuclear Physics (USSR)*, **43**, 304.

AUTHOR INDEX

Abrikosov, A. A. 7–10
Amado, R. D. 38–42, 136, 150
Aniol, K. 40
Anthony, I. 145–7
Antonov, A. N. 27–8, 31, 32, 87, 114–17, 120, 128, 130–3, 135–9, 140–4
Araseki, H. 41, 141–2
Avan, M. 40, 141–2
Azhgirey, L. S. 41

Baldit, A. 40, 141–2
Banerjee, M. K. 68–9, 88, 121
Baranger, M. 102
Barrett, R. C. 145
Bartel, J. 25
Bartz, B. I. 55
Beiner, M. 131
Bellicard, J. B. 109–10, 112, 118
Belyaev, S. N. 146–7
Belyakov, V. A. 47, 48, 61, 70
Benhar, O. 86–7, 115
Bernheim, M. 40, 142–3, 145, 147–8
Bertsch, G. F. 41
Bertozzi, W. 104, 106, 145, 148
Betschek, A. G. 88
Bethe, H. 23, 66, 88
Blanpied, C. 40
Blin, A. H. 63
Boal, D. H. 40–1
Bodek, A. 40
Boffi, S. 143, 145, 148, 150
Bohigas, O. 25, 50, 65, 77–82, 118, 129
Bolotin, Yu. L. 55
Bortignon, P. F. 52, 54, 66
Brack, M. 25
Brandenberg, R. A. 86
Brandow, B. H. 88
Branford, D. 145–7
Brink, D. M. 56–7, 88, 128
Broglia, R. A. 52, 54, 66
Brown, B. A. 113
Brown, G. E. 88, 111, 139
Brueckner, K. A. 24, 32, 34, 37, 66, 68, 89, 139

Buchler, J. R. 24, 139
Buck, B. 64
Burov, V. V. 32, 39
Buss, W. 148
Bussiere, A. 40, 142–3, 145, 147–8

Calogero, F. 41–2
Camis, R. Mc. 40
Campi, X. 25, 57, 106, 110
Capitani, G. P. 40, 145, 148
Capuzzi, F. 143
Casas, M. 122–4, 136
Castor, J. 40, 141–2
Cavedon, J. M. 109–10, 112, 118–19
Chaigne, G. 40, 141–2
Chandra, H. 104, 106
Christov, Chr. V. 128, 130–3, 136–7
Ciofi Degli Atti, C. 86–7, 95, 115, 148
Clark, J. W. 65, 77–8, 87, 118
Clark, R. C. 24
Clement, C. F. 102–3
Cordell, K. R. 40–1, 141–2
Cotanch, S. R. 149
Czyż, W. 35, 47, 70

Da Providencia, J. 88–9
Dal Rì, M. 50, 79, 81–2
Dasso, C. H. 52, 54, 66
De Forest, T. 86, 106
De Nercy, A. 142–3, 147
de Sanctis, E. 40, 145, 148
Deb, B. M. 26–7
Dechargè, J. 57, 64, 116, 118
Degasperis, A. 41–2
Dellagiacoma, F. 88
Dennis, L. C. 40–1, 141–2
Devaux, A. 40, 141–2
Didelez, J. P. 40, 141–2
Dieperink, A. E. L. 86
Dirac, P. A. M. 3
Doering, R. R. 40–1, 141–2
Durand, M. 25

Durso, J. W. 111, 139
Dzyaloshinsky, I. E. 7–8, 10

Eden, R. J. 34, 37, 89
Eisenberg, J. M. 150
Eldyshev, Yu. N. 32
Elliot, J. P. 105
Elton, L. R. B. 87, 102, 147–9
Emrich, K. 70
Engelbrecht, C. A. 143
Escudero, J. I. 113
Ey, W. 58, 70, 72–4, 80, 86, 88, 92, 95, 121, 130–1, 136, 148

Fantoni, S. 65, 78, 83–5, 87, 89, 93, 95, 118, 120
Fargeix, J. 40, 141–2
Fermi, E. 22
Findlay, D. J. S. 145–8
Flocard, H. 128, 131
Flowers, A. G. 145–7
Flynn, M. F. 65, 77–8, 118
Force, P. 40, 141–2
Francis, N. C. 34, 37, 89
Frankel, S. 38, 40, 141
Frati, W. 38, 40
French, J. B. 103
Friar, J. L. 104, 106, 110
Fricke, G. 137–8
Frois, B. 109–10, 112, 118–19
Frullani, S. 40, 142, 145–9
Fujita, T. 40–1, 141–2

Galitski, V. M. 47
Gamba, S. 105
Gardiner, S. N. 147–8
Gari, M. 70, 74, 146
Gaudin, M. 76–7, 81, 83, 118, 121
Ghosh, S. K. 26–7, 63
Gillebert, A. 142–3, 147
Gillespie, J. 76–7, 81, 83, 118, 121
Girod, M. 57
Giusti, C. 143, 145, 148, 150
Gogny, D. 57, 64, 109, 118
Goldstein, V. A. 145–6
Gonchar, V. Yu. 55
Gorbunov, A. N. 145–6
Gorkov, L. P. 7–8, 10
Göringer, H. 146–9
Gottfried, K. 35, 47, 70
Goutte, D. 118–19
Grashin, A. F. 41, 141–2
Greeniaus, L. G. 40
Griffin, J. J. 126, 134
Gross, D. H. E 143

Grypeos, M. E. 88
Guet, C. 25
Gupta, U. 27
Gurvitz, S. A. 40, 141–2

Håkansson, H. 25
Hansen, A. O. 145–6
Hasse, R. W. 63–5
Hatch, R. L. 40
Hebach, H. 146
Heisenberg, J. 104, 106
Highland, V. 38
Hill, D. L. 126
Hiller, B. 41, 63
Hiramatsu, S. 142
Ho, P. Z. 109, 110, 112, 118
Ho Xuan Phan 142–3, 147
Hodgson, P. E. 28, 32, 87, 100–2, 104, 106, 108, 110–17, 136–9
Hoffmann, G. W. 40
Hohenberg, P. 26
Huber, M. G. 145
Huet, M. 109–10, 112, 118–19
Hüfner, J. 31, 41, 123
Hugenholtz, N. M. 14
Hutcheon, D. A. 40
Hyuga, M. 70, 74

Ignatenko, M. A. 41
Inopin, E. V. 41, 55
Ivanov, V. V. 41
Izutsu, M. 142

Jackson, D. F. 145
Jaminon, M. 59, 114–16, 118–20, 124, 141–2
Jans, E. 40
Jastrow, R. 75, 89
Jeukenne, J. P. 52, 60, 61, 66
Jibuti, R. I. 145–6
Johnson, M. B. 111, 139
Jorna, S. 24, 139

Kamae, T. 142
Kasatkin, Yu. A. 41
Kennet, T. J. 146–7
Keserashvili, R. Ya. 145–6
Khodel, V. A. 125
Kiergan, S. E. 145–6
Kirichenko, I. K. 41
Kirzhnitz, D. A. 25
Kobe, D. H. 16, 59
Koester, L. J. Jr. 145–6

AUTHOR INDEX

Köhler, H. S. 66, 143
Kohn, W. 26
Komarov, V. I. 40
Koonin, S. E. 40
Kosarov, G. E. 40
Kozin, A. B. 146–7
Kozlowski, T. 40
Krivine, H. 25, 125–6, 129
Kubodera, K. 141–2
Kümmel, H. 70
Kuo, T. T. S. 88
Kuplennikov, E. L. 145–6
Kutzelnigg, W. 17, 59
Kuznetzov, A. S. 41

Lagaris, I. E. 84
Landau, L. D. 140
Landau, R. H. 41, 140
Landdaud, C. 40, 141–2
Leconte, P. 109–10, 112, 118–19
Lee, S. Y. 88
Leitch, M. J. 145–6, 148–9
Lejeune, A. 52, 60, 61, 66
Levinger, J. S. 146, 149
Lifshitz, M. 41, 140
Liljestrand, R. L. 40
Lin, F. C. 145
Lipkin, H. J. 105
Lipperheide, R. 143
Liuti, S. 86–7, 115
Lombard, R. J. 24, 25, 32, 139
Löwdin, P.-O. 3, 98
Ludeau, P. 110, 112, 118
Lührmann, K. H. 70
Luhrs, G. 146
Lukyanov, V. K. 32, 39
Luttinger, J. M. 64

McGuire, J. B. 41
Madurga, G. 113
Mahaux, C. 48, 52, 54, 59–61, 66, 68–9, 114–16, 118–20, 124, 141–3
Malaguti, F. 100–2, 104, 106, 108–12, 114, 117, 126, 136–7, 139
Malecki, A. 51, 89
March, N. H. 32
Martorell, J. 106, 110, 122–4, 136
Massen, S. E. 113
Matthews, J. L. 145–8
McGeorge, J. C. 145–7
McGuire, J. B. 41
Merle, K. 137–8
Mescheryakov, M. G. 41
Migdal, A. B. 14, 47, 125
Millener, D. J. 102

Miller, H. G. 148
Miller, L. D. 39
Miramatsu, H. 142
Miska, H. 137–8
Morris, C. 40
Moss, G. A. 40
Moszkowski, S. 88
Mougey, J. 40, 142–3, 145–9
Moya de Guerra, E. 122–4, 136
Müller, H. 40

Nagorny, S. I. 41
Nakada, A. 110, 112, 118
Nakamura, K. 142
Narodetsky, I. M. 43
Nechkin, A. A. 146–7
Negele, J. W. 56–7, 87, 104, 106, 110, 148
Nemes, M. C. 31, 41, 123
Netzband, D. 40
Ngô, H. 59, 114–16, 118–20, 124, 141–2
Nikolaev, V. A. 28, 31, 87, 115, 130–1, 135–7, 140–4
Noble, J. V. 150

Orland, H. 62–5, 143
Orlandini, G. 86–8, 90, 92–5, 114–15
Owens, R. O. 145–8

Pacati, F. D. 143, 145
Pacati, F. F. 150
Pace, E. 95
Padjen, R. 64
Pandharipande, V. R. 65, 83–6, 89, 93, 95, 116, 118, 120, 136
Panoff, R. M. 65, 77–8, 118
Papanicolas, C. N. 116, 118–19
Parks, R. L. 40–1, 141–2
Perey, F. 64, 103, 107
Peridier, C. A. 145, 148
Petkov, I. Z. 27–8, 31, 32, 87, 114–17, 120, 128, 130–3, 135–9, 140–4
Phan, X. H. 118–19
Picchi, P. 51, 89
Platchkov, S. 109,
Platchkov, S. K. 118–19
Pol', Yu. S. 32
Prestwich, W. V. 146–7
Priou, M. 142–3

Quentin, P. 131

Rajagopal, A. K. 27

Ratcliff, K. F. 88
Rawlins, J. A. 148
Razin, S. V. 41
Reide, F. 40, 141–2
Reuter, W. 137–8
Ricco, G. 105
Ridge, R. 40
Ring, P. 25, 64
Ripka, G. 76–7, 81, 83, 118, 121
Ristig, M. L. 78
Ritchie, J. L. 40
Roberts, B. L. 145–6, 149
Roche, C. 40, 141–2
Rosati, S. 78, 84, 87
Rothhaas, H. 110
Roy, G. 40
Royer, D. 40, 142–3, 148

Salmè, G. 86–7, 95, 115
Saperstein, E. E. 125
Sapp, W. W. 145, 146
Sargent, C. P. 145, 146, 148
Sartor, R. 47–8, 68–9, 128, 143, 145
Sauer, G. 104, 106
Schaeffer, R. 62–5
Schäfer, L. 101
Schiavilla, R. 85–6, 136
Schier, H. 146, 148
Schoch, B. 146–9
Schuck, P. 25, 63–5
Schweizer, T. C. 40–1, 141–2
Scott, B. 88
Semyonov, V. A. 146–7
Senè, M. R. 145–7
Shakin, C. M. 88–9
Shalamov, Y. Y. 41, 141–2
Sham, L. J. 26
Shotter, A. C. 145–7
Sick, I. 86, 108–10, 112, 118, 142–3, 147
Siemens, P. J. 24
Simonov, Yu. A. 43
Singer, P. 41
Sips, L. 116, 118
Skyrme, T. H. R. 105
Smith, V. H. Jr 17, 59
Sprung, D. W. L. 57, 66, 106, 110
Stetz, A. W. 40
Stiehler, T. 40
Stoletov, G. D. 41
Stringari, S. 50, 65, 77–82, 118
Svennve, J. P. 55
Swift, A. 87, 102, 147–9

Tang, Y. C. 127
Tarnowski, D. 40, 145, 148
Tesch, S. 40

Thiessen, H. A. 40
Thomas, L. H. 22
Thorley, P. J. 145–7
Thornton, S. T. 40–1, 141–2
Thouless, D. J. 6, 22, 88
Titov, A. I. 39
Traini, M. 86–8, 90, 92–5, 114–15
Treiner, J. 25, 129
Truex, W. 68–9, 88, 121
Tsai, J.-S. 146–7
Turchinetz, W. 145, 148
Turck-Chieze, S. 40, 145, 148

Uguzzoni, A. 100–1, 104, 106, 108–12, 114–15, 117, 136–7, 139

Van Dyck, O. 38, 40
Van Giat, N. 131
Van Orden, J. W. 68–9, 88, 121
Vasilyev, O. V. 146–7
Vautherin, D. 56–7, 128
Verondini, E. 100–1, 104, 106, 108–12, 114–15, 117, 136–7, 139
Vicente, J. 40, 141–2
Villars, F. 88
Viñas, J. 113
Vzorov, I. K. 41

Wagner, G. J. 142–3, 145, 147–8
Walecka, J. D. 106
Wambach, J. 116, 118
Watase, Y. 142
Webb, S. J. 102
Weber, M. J. 39, 150
Weidenmüller, H. A. 101, 143
Weise, W. 145
Werbeck, R. 38
Wheeler, J. A. 126, 134
Whitten, C. 40
Wildenmuth, K. 127
Williamson, S. E. 118–19
Willis, A. 40
Willis, N. 40
Winter, J. 63
Wiringa, R. B. 85–6, 136
Woloshyn, R. M. 38–42, 150
Wong, C. W. 126
Wood, S. A. 145, 146
Wortberg, A. 146

Yang, D. 40

Zabolitzky, J. G. 58, 70, 72–4, 80, 86, 88, 92, 95, 121, 130–1, 136, 148
Zimmerman, C. H. 145–7
Zhmyrov, V. N. 41
Zverev, M. V. 125

SUBJECT INDEX

Ag 38
annihilation operator 8, 53
atom (many-electron) 23
auxiliary potential 61, 66

Be 38
Bethe–Faddeev equation 71
Bethe–Goldstone equation 66, 71
^{209}Bi 147
binding energy 26, 33, 55, 59, 85
breathing vibrations 128, 131–2, 139–40
Brueckner G-matrix 68
Brueckner–Hartree–Fock theory 54, 65–70, 96
Brueckner orbital 73
Brueckner method 35–37, 63, 88, 143

^{12}C 38, 40, 86, 138, 147
^{40}Ca 32–3, 56–8, 63, 81, 86, 90–3, 95, 108, 109, 114–17, 121, 125, 136–7, 140–2, 144
^{48}Ca 108–9, 125, 136
centre-of-mass correction 103, 105
centroid energy 102
charge density difference 109
charge distribution (nuclear) 56–7, 96, 118–19, 136, 139
clusters (nucleon) 39–40, 77–8, 85, 87
coherent density fluctuation model 28, 87, 134–41
collective excitations 24
commutation relations 7–8
core polarization 61–2, 110
correlation
   contribution 64–5
   dynamical 58
   function 75–6, 78
   graph 62
   nucleon–nucleon 37, 146
   operator 89
   Pauli 58

Coulomb
   energy 55
   potential 23
   terms 24
coupled-cluster theory 70
creation operator 8
^{133}Cs 147
Cu 38

Darwin–Foldy term 106
deep states 107, 143
degeneracy (spin–isospin) 84
delta forces 41, 131, 136
density distributions (nuclear) 1–150
   charge 56–7, 104, 112–13, 118–19, 136, 139
   definition 19–22
   non-monotonic 32
   neutron 111, 113
density functional theory 26
density dependence 54, 122, 124, 148
density matrix
   one-body 3–7, 15–17, 21–2, 26, 45, 49, 51, 71, 76–7, 79, 96–7, 121, 134–5
   two-body 6–9, 76–7
de Tourreil–Sprung super soft-core potential 68, 72, 74
dispersion relations 14, 64
distorted wave impulse approximation 143, 145–6
Dyson equation 12, 52

(e, e'p) reaction v, 34, 142–50
effective interaction 24
energy
   functional 25
   kinetic 21–2
   total 23, 26
energy-density formalism 24, 27
energy-density functional 24, 29
Euler–Lagrange equation 24, 29, 76
exp(S) method 70–4, 95–6, 130, 136, 148

F 40
Fermi
  energy 59, 103
  gas 31, 35, 44–8, 68–9, 122–4, 136
  hypernetted chain 78
  liquid 47
  momentum 14, 22, 39, 45
  sea 16, 62, 114, 116–18, 120
  sphere 46
few-body system vi
field operator 7
final state interaction 40
finite size effects 122
flucton 39
fluctuation model 28
Fourier transform v, 12, 27, 42, 49, 55, 120, 129

gamma reactions 145–50
generator co-ordinate method 126–34, 136
giant multipole resonances 34
G-matrix expansion 54, 68
Gogny interaction 64, 65
Gram–Schmidt method 106
Green function method 10–15, 21, 51–3, 62, 125, 143
  one-particle 10
  two-particle 11
Griffin–Hill–Wheeler generator co-ordinate wave equation 127–9

Hamada–Johnston potential 74, 148
hard core 35, 47
harmonic oscillator model 48, 50, 68, 79, 83, 86, 89, 105, 115, 129–33, 148
Hartree approximation 42
Hartree–Fock theory v, vi, 12, 16, 25, 44, 51–74, 96–7, 101, 109, 111–12, 115, 119, 122, 124, 129–30, 148
^{3}He 41, 86
^{4}He 50–1, 58, 72, 79–82, 86, 90–5, 129–33, 136, 145–6
Heisenberg
  operator 9
  representation 9–10
Hohenberg–Kohn theorem vi, 26–33
hole-line expansion 60, 65–70
hypernetted chain (Fermi) 78

incompressibility 140
independent-particle model 44–59
inelastic scattering (proton) 38

Jastrow method 75–96, 121, 145, 148

^{39}K 108, 136
kinetic energy 21–2, 24
knock-out reaction 145, 149

Lagrange multiplier 27, 30, 32
Lagrange quasi-particle method 125
Lehmann representation 53
^{7}Li 145
local density formalism 19–33

magnetic moment 147
many-body problem v, 1
many-electron atom 23
mass operator 13, 60–70
mean field calculations 122
meson exchange currents 146
momentum distribution (nuclear) 1–150
  theoretical basis 1–18
monopole vibration 128, 131–2, 139–40
Morse potential 140
Moshinsky transformation 89
multiparticle–multihole configuration 63, 101
muonic atom 136

Na 40
natural occupation numbers 98
natural orbital representation 15–18, 96–121
Ne 40
^{58}Ni 107–8, 142, 144
non-local potential 103, 107
nuclear matter 85, 93, 135, 139
  wavefunction 83
nucleon–nucleon
  correlation 37, 41, 75, 88–95
  interaction 2, 33, 35, 47, 58–9, 62, 74, 101, 138

^{16}O 32–3, 41, 51, 56–8, 69, 73, 86, 87, 90–3, 95, 129–30, 132–3, 136, 141–2, 145, 147–8
occupation number representation 8
occupation probabilities 64, 96–121
one-body density matrix 3–7, 15–17, 21–2, 26, 45, 49, 59, 71, 76–7, 79, 96–7, 121, 134–5
optical potential 64
orthogonalization (of wave function) 106
oscillator states 71

pair correlation function 34
particle–hole excitation 70, 101

SUBJECT INDEX

Pauli
  correlation 58
  principle 52, 97
  operator 68
^{206}Pb 107
^{208}Pb 40, 56–7, 110, 113, 115, 119, 125, 137, 141–2, 147
Perey factor 103, 107
perturbation expansion 61, 66, 76
phase shift analysis 33
photon absorption 146, 149–50
pick-up reactions 102–3
pion production v, 39
pion reactions 41, 141–2
plane wave impulse approximation 145–6, 148
(p, n) charge-exchange reaction 149
polarization contribution 64–5
pole representation 13
potential energy 30, 32
(p, 2p) reaction v, 34, 107
Pt 38

quasi-deuteron model 146
quasi-particle states 14

radii (nuclear) 55, 59, 102, 104–5, 133
random phase approximation (RPA) 63, 97, 111–12, 116, 119, 146
reaction matrix 66
Reid soft-core potential 58, 68–9, 72, 74, 87, 92, 131
relativistic corrections 105–6
repulsive core 76
Ritz variational principle 127
RPA 63, 97, 111–12, 116, 119, 146

Saxon–Woods potential 48, 68, 79, 102, 105, 114, 118
Schrödinger
  operator 9
  representation 9–10
second quantization 7–10
semi-classical model 63
separation energy 32–3
shell model 35, 48–51, 57, 81
short-range correlations v, 34, 75–97, 116–17, 121, 131–2, 137–8, 145
^{28}Si 142–4

single-particle potential v, 64, 96–121, 136–7
single-particle states 71
Skyrme effective interaction 25, 55–7, 111, 122, 128–9, 131
Slater determinant 17, 44, 49, 54, 74, 80–1, 89, 97, 100, 115, 117, 121, 128
^{120}Sn 122–3
^{132}Sn 125
spectral function 14, 142–4
spherical potential 48
spin–isospin degeneracy 84
spin–orbit correlation 83
spin–orbit energy 55
SPP method 96–121
Sprung potential 68–9, 72, 74
square well potential 131–3
stripping reactions 102
Strutinsky method 25

Ta 38, 141–2, 147
tensor correlations 83, 86–97, 115–17, 121
Thomas–Fermi model 22–5, 31, 63
three-nucleon interaction 85
total energy 23, 26
two-body
  density matrix 6–9, 76–7
  operator 2
two-step mechanism 149

unitary transformation 98

variational
  calculation 83–8
  equation 24, 32
  principle 29, 127
velocity-dependent interaction 54

Weizsäcker correction 25
Wigner
  distribution function 21, 122–4, 135
  force 76
  space 64
  wave function 125
wound parameter 78

zirconium isotopes 109–11
^{90}Zr 56–7, 110, 122–3, 125, 137, 139